Writing Style and Communication Skills for Undergraduate Engineers

College Publishing books are printed on acid-free paper.

ISBN: 978-1-932780-25-3

(This title is also available as an e-book purchase (e-book ISBN: 978-1-932780-26-0) at http://www.vitalsource.com or http://www.redshelf.com through your university bookstore.)

College Publishing
12309 Lynwood Drive, Glen Allen, Virginia 23059
Phone (804) 364–8410
Fax (804) 364–8408
Email collegepub@mindspring.com
Internet http://www.collegepublishing.us

Writing Style and Communication Skills for Undergraduate Engineers
Fifth Edition

by Katie C. Russell,
Jeffrey Donnell and
Sheldon Jeter

College Publishing
Glen Allen, Virginia

ABOUT THIS GUIDE

Writing Style and Communication Skills for Undergraduate Engineers, Fifth Edition, address-es two primary questions of inexperienced technical writers: "How should my report be writ-ten?" and "How do I present my experimental work appropriately using figures, tables, equa-tions and spreadsheets?" The first two sections of this guide respond to these two questions.

The first section of this guide, *Writing Style in Undergraduate Engineering Reports,* speaks to the problems that students face as they draft their reports. Specifically, it outlines the norms of format in engineering reports, and it describes the way format is linked to the sub-stance of the report. This section of the guide also provides a set of guides and example reports based on the writing requirements of the different laboratory courses at the Georgia Institute of Technology. It also includes an example of a report that would be written as part of an undergraduate research experience. These model reports are offered as examples of good reports, and they should be consulted when questions arise concerning format, style, or pre-sentation of data. *Writing Style and Communication Skills for Undergraduate Engineers* also provides guidance on effective oral and visual communication with attention to slide presen-tations, posters, and tips on speaking effectively.

The second section of this guide, *Standards for Undergraduate Engineering Reports,* speaks to the problems students face as they present their work using graphical elements. This section of the guide explicitly outlines the norms for assembling and labeling figural and tabular infor-mation, provides examples of well-made figures and tables, and demonstrates how to integrate graphical elements into written documents. Detailed checklists help students determine whether their data are professionally presented. This section guides students through navigating resources such as the engineering library and online research databases. *Standards for Undergraduate Engineering Reports* also reviews the norms of paragraph and sentence formation and other methods to achieve clear and logical writing in technical reports. New to this Fifth Edition is a chapter devoted to online courses and virtual meetings, with particular attention to Zoom and Microsoft Teams.

The new third section of this guide, *Writing to Land the Job,* offers step-by-step guidance to students on how to successfully navigate the job search and application process, from resume writing to interviewing skills. Several sample resumes are provided as well as effective means to communicate with prospective employers. For students interested in graduate school, this sec-tion provides guidance on how to effectively identify programs of interest and apply to them. Platforms, like LinkedIn, that have become essential in the job search and candidate presentation process are fully explored.

The fourth section of this guide, *Writing on the Job,* speaks to the kinds of tasks students face when they make the transition from classroom reporting to workplace communication, where problems are often open-ended and audiences cannot be assumed to be engineering professionals. This section of the guide concentrates on a design report and presentation for an open-ended project. This report is prepared to describe a technical project to non-technical readers, and this section highlights the visual and verbal steps the author has taken to accommodate these non-technical readers.

The Publisher would like to gratefully acknowledge the work of Professor Katie C. Russell, Senior Professor of Practice at Tulane University's School of Science and Engineering, to make the Fifth Edition a reality. Professor Russell's new Chapter 2.18, *Virtual Communication Platforms*, covering Zoom and Microsoft Teams, as well as the new Part III, *Writing to Land the Job*, covering resume writing, interviewing and LinkedIn, has made this Fifth Edition an invaluable addition to engineering students' collection of essential guidebooks.

Table of Contents

PART THREE
Writing to Land the Job

PART FOUR
Writing on the Job

Part One:
Writing Style in Undergraduate Engineering Reports

CHAPTER 1.1

GLOBAL STYLE: AN INTRODUCTION TO UNDERGRADUATE ENGINEERING REPORTS

What are technical reports for? Who reads them, and what do readers do after they read them? What should a report show about your work? These questions about reports must be answered by talking about the people who read them, and you need to have answers to these questions before you write. Here, we show you how to answer these questions. We explain what technical reports are for and what readers need to do with your reports. We explain the conventions for organizing reports, and we explain how to make your reports easy to read. In Section 2 of this book, we also explain the conventions for presenting your data visually.

What is technical writing for?

Your goal in a technical report is to present data that you have collected during a project and to account for the actions you took to obtain that data and the actions you took to record and preserve the measured values. Your data will usually include direct measurements and calculated values based on those measurements. Your reports will present that data using some combination of experimental diagrams, photographs, data tables, plots that compare predicted and measured values, and other kinds of displays, depending on the project and your equipment.

Your job as you write your report is to explain fully all the actions you took to collect and analyze your data, because the way you took the measurements and the way you display the results always influence the reliability of your work. In order for readers to fully understand the system you measured, they have to know all of your actions in collecting, analyzing and displaying data about that system. You are also expected to draw reasonable conclusions based on your results, and possibly to make relevant recommendations to improve operation, efficiency, safety, and so on.

The goals of technical reports

People use technical reports to meet different kinds of goals. Your supervisors and colleagues may use your reports to characterize systems, to demonstrate that components meet performance specifications, to diagnose problems, or to verify solution methods. However, your technical reports may remain in your company's files for a long time. Your old reports might be read by engineers who wish to maintain or modify a system that you worked on, or by people who are new to your project and who need to become familiar with it. If there is litigation concerning a system that you worked on, your reports will likely be reviewed by attorneys or their technical experts. It is important that all such readers fully understand what you did, what conclusions you drew and what recommendations you made.

When you prepare a report, you should strive to accommodate all of these possible readers. In most cases, you can do this by preparing reports that are simple, accurate, and accessible. You can make your report seem simple by writing an Introduction that fully defines the problem or question that the project addressed and the result that was obtained. You assure that a report is accurate when your apparatus and procedures descriptions account for all the instruments you used and all the steps you took to calibrate the system, then to collect, store and analyze the data. You make your report accessible first by formatting your displays according to professional norms, and second by describing them fully in the text of the report.

How format makes reports accessible

In order to conduct an experiment, you must do several things in order. You must determine what you wish to measure, you must obtain and calibrate the right equipment, collect your data, analyze that data, and then determine what it means. When you prepare your project report, you will subdivide it into sections that describe each of these project steps in logical order. Each of these sections should have a boldface heading. The sequence of headed sections in a report is called the report's *format*.

Professionals use the term *format* to describe the headings and the flow of information in a report. In most reports, the headings you will use and the information you will provide are these:

- Introduction sections explain what needs to be measured and why
- Apparatus and Procedures sections define what equipment was used and what calibration, measurement and analysis steps were taken
- Results sections display and describe the measured and calculated results
- Discussion sections explain what is significant about the data and comment on any possible sources of error in the work
- Conclusion sections briefly summarize the main points made in each of the previous headed sections.

When you subdivide your report into these sections, you will create an orderly explanation that displays the logic of your work. When you describe all of the steps of data collection and analysis, your report will be complete, specific and easy for any reader to understand.

Responsibility, transparency and cover sheets

Reports are prepared and delivered as part of transactions between people, such as supervisors and employees, consultants and clients. These people have various kinds of jobs with various kinds of responsibilities to their companies and colleagues. When you submit a report, you and your team accept responsibility for all of the work and conclusions the report describes. You submit your report to a person who acts as your supervisor or instructor; that person accepts responsibility for assigning your work and for approving any research or investigation you may have done. When you submit a report, you must explicitly display the professional relationships that drove your work. Usually you will do this by preparing a complete cover sheet.

When students prepare lab reports for their project teams, they are usually asked to list their course section ID, their report's title, the date of submission, the instructor's name, and the names of all contributing team members. This information makes it easy to sort papers for grading, but it also fully defines the lines of responsibility for the report, as well as the date of completion and the complete title of the work. Such classroom cover sheets do not differ greatly from the cover sheets that you will prepare when you are in practice as a professional.

This book presents two variant report forms—letter reports and memorandum reports that may be used in certain professional circumstances. You should notice that these report forms vary primarily in the ways that responsibilities are displayed on the first page of the report. In both of these alternative report forms, the authors/team members are identified, the supervisor/recipient is identified, and the report's title and submission date are presented. While the arrangement or page layout of this information can vary from organization to organization, the need for you to display the names of responsible parties will not change.

Completeness in technical reports

Readers expect technical reports to be complete. A complete report documents an experiment thoroughly, describing calibration actions, data collection and analysis, with methods descriptions and results displays for each stage of the project. When a report completely documents a project, the results will be easy to reproduce, and the explanations will be easy to understand.

Technical reports are often long because projects have many steps to document. Experimental apparatus is often elaborate, calibration procedures can be time consuming, and data may require several steps of analysis. You should document each experimental step in a notebook, including calibration, data collection and storage, and analysis, and then draw on the notebook records when you compile the report itself. If you have kept thorough records in a notebook, you should find that much of your report can be transcribed from those records.

Reports in the classroom and in the real world

The reports you prepare in your classroom will probably differ somewhat from those you prepare when you join an engineering firm. This is to be expected. It is your instructor's job to verify that you have learned to keep complete project records and to verify that you have prepared fully detailed reports. We know that our laboratory classes are the only places where you will learn to do this. When you enter practice, you should talk with your supervisor to determine what level of detail is required in your reports, and you should follow that guidance. However, while reporting requirements may change after you enter practice, your record-keeping practices should not change; you should continue throughout your career to keep detailed records of your project work, your raw data and your analysis steps.

Expectations for data presentation

In your reports, you will present data visually, using, for example, tables, plots and photographs. In Part 2 we will provide some guidelines for preparing high quality data displays. Many readers will be able to understand your work simply by examining your displays. However, many readers will not be sufficiently familiar with your project to understand your results, their importance, or the way you quantify error. When you display results in a report, you need to describe each display, explicitly describe the significance and the error, and then name any other considerations that readers need to take away from the display. All important considerations must be presented fully in your reports; if readers are not able to find important warnings or caveats in the pages of your report, you may be blamed later for misjudgments that your readers make.

Describing your displays

When you conduct an experiment, you are looking for something, and the experiment is designed to help you find it. In introductory projects, you usually look for a trend or for a peak value, and you then compare your measured values to a model that defines ideal or expected values. When you prepare and describe your displays, you should state how well your data compares with your expectations or projections. Such statements need not be elaborate, but they should be specific, as is this example:

> The profile of normalized velocity is uniform, with a decrease *of 5% at .9 normalized radius*. This differs greatly from the prediticted profile, where the normalized velocity varies continuously, *decreasing to 0.4% at .9 normalized radius*.

In this example, the first sentence describes the experimental data that you may have collected, while the second sentence describes predicted or calculated values. The phrase "This differs greatly from the predicted profile" is your statement of the primary trend or observation that the data represents.

The italicized terms here demonstrate how this example avoids vagueness by using numbers to exactly quantify the velocity decreases at 5% and 0.4% for the experimental value and the predicted value. When presenting results or making comparisons, inexperienced writers often use unacceptable terms, such as "around," or "close to." Such vague terms show up in this statement:

> The data agree *reasonably well* with the model.

To avoid vague terms, always use numbers to describe, evaluate or compare your data. When numbers are used, the vague statement above becomes:

> The data agreed with the model *within +/– 10%*.

In addition to making points about your data, you must describe your displays to identify all the things your reader should see. For plots, you should define what each type of marker represents, as well as trendlines, models and calculations that you have imposed on the display. For table displays, you should name the main categories represented by the rows and columns and indicate how calculated values were derived. For diagrams and photographs, you should use labels to call attention to features that readers need to see, and you should list and comment on those features in a text description adjacent to the display.

Discussing data

After you have described your displays, you should describe and explain any anomalies or irregularities that may be present in your data. It is expected that student lab data will differ from predictions that assume ideal conditions; however, you are still expected to quantify your data's divergence from predictions, and you are expected to explain why your model and your data disagree. Further, your data will occasionally appear to be outrageously wrong or even physically impossible. Under these circumstances, you need to revisit your data collection methods in order to establish that the unusual result was (or was not) the result of a setup error or a calculation error. In all cases, your discussion of data and anomalies should be grounded in a scientific understanding of the system under study and the equipment used to conduct the study.

Erroneous results can sometimes be traced to data-entry errors. This is common on projects that involve data records that are handwritten or typed on the fly. When data entry errors occur, it is acceptable to exclude the erroneous data points from your calculations provided that you do three things:

1. Consult with your instructor or TA who will approve any exclusions from your data set.
2. Preserve the data table with the erroneous values displayed. In that table, you should indicate which values are to be excluded from your calculations by annotating or footnoting the data table.
3. Disclose and justify your exclusion of data points as part of your discussion of the display.

Details make your data credible

In order for your results to be credible and valid, you must define what equipment and methods were used, and you must present numbers professionally. When you describe experimental apparatus, you should provide the manufacturer and model number for each item of equipment, and you should define any specific settings that were used. When you display numerical results, your numbers need to be reasonable as well as accurate. Your calculators or spreadsheets will, for example, allow you to calculate a velocity of "2.141492654 m/sec," but the trailing digits in this number almost certainly exceed the resolution of your instruments. To display calculated values responsibly, you should review your instructor's guidance on significant digits in calculations. You may also wish to review the discussion of significant digits in Part II of this book.

Headings and information in reports

We mentioned earlier that the term *format* is used to describe the sequence of headed sections in a report. Most experimental reports are divided into sections using these headings:

1. **Introduction**
2. **Methods**
3. **Results**
4. **Analysis**
5. **Discussion**
6. **Conclusions**

When your headings are arranged in this order, the running text of your report will naturally display the logic of your project. In order to do this, however, you must be sure that you provide the right kind of information in each of section of the report. Here we briefly outline the kind of information you are expected to present in each of these main sections of a report.

Introduction

The *Introduction* is the first section of a technical report. For narrow classroom projects, instructors may use the words *Objectives* or *Background* to name this opening section; we use the term **Introduction** because it can be applied across a wide variety of report types.

In an Introduction section, you should:

1. Define the <u>need</u> or <u>problem</u> or <u>question</u> that the project addresses,
2. Describe <u>related work</u> that you are aware of,
3. State the particular <u>goal</u> that the project addresses, and
4. State the measurements and calculations that you will perform as you pursue that goal.

Undergraduate experimental projects may be highly scripted for the students; reports on such projects sometimes abbreviate the Introduction section by omitting the need and related work statements. These statements are typically required, however, for more open-ended capstone projects.

If you have trouble writing the Introduction section of a report, it is best to begin by answering these questions:

1. What need does this work address? Why should people care about this work?
2. Have others done work that is pertinent to our work on this project?
3. What problem does this work solve? (or What question does this work address?)
4. What measurement and analysis steps did we take in order to solve the problem?

Apparatus and Procedures

Apparatus and Procedures, often simply called **Methods**, is the second section in most undergraduate reports. In the Apparatus and Procedures section, you should

1. Describe the instruments that you used to collect data,
2. Describe what particular measurements you took
3. Describe how you recorded and preserved your data for later use.

For most projects, you will use commercially available instruments. In these circumstances, your apparatus descriptions should include the instrument's manufacturer and model number, as this enables readers to exactly reproduce your work. The itemized report in Chapter 1.3, for example, required students to use a function generator; in the report, that instrument was fully described as follows:

> The voltage was generated by a *Gwinstek SFG-2100 function generator*

The italicized material adequately characterizes the apparatus by defining maker and model.

For some projects, you might use non-commercial instruments that were fabricated in-house. In these cases, you should provide a functional description of the instrument that includes any specifications that may be critical to your work. The itemized report of Chapter 1.3 describes a department-fabricated amplifier in this way:

> Output signals were processed via a department-fabricated, non-inverting, constant gain amplifier

If you have trouble writing the Apparatus and Procedures section of a report, it is best to begin by answering these questions

1. What instruments were used for data collection?
2. What measurements were obtained?
3. Where was the data stored?

Results

The **Results** section, alternatively called **Data** or **Findings**, is usually the third section of an experimental report. In the Results section you should

1. Display your results,
2. Describe your results,
3. Account for discrepancies and anomalies in your reported values.

For undergraduate experimental projects, the Results section is typically the largest part of a project report, as it is the section where most data displays are placed. Individual experimental results are generally presented as data displays with accompanying text descriptions, which are discussed in the section *Describing Your Displays*. However, Results sections commonly assemble the data from numerous measurements and calculations; as you assemble your reports, it is important for you to review your class instructions to verify that your report includes all of the required displays.

When you have assembled all of your displays in the appropriate order, you should review the text descriptions to assure that they include this information:

1. What measurement does the display represent?
2. What instruments and methods were used to obtain the data?
3. What analysis steps (if any) were used to process the data in the display?
   Mention key equations or methods used in the analysis.
4. What trend or characteristic does the display demonstrate?
5. What features of the display represent that trend or characteristic?
6. What errors or anomalies does the data contain?

Discussion

The **Discussion** section of a report is placed after the results presentation. In the Discussion section, you should explicitly state and support any conclusions that can be drawn from the data that you have presented. In the **Discussion** section, you should usually speak to these issues:

1. What is the logical interpretation of these results; what do the trends imply?
2. How well do the experimental results compare with predicted values?
3. How can discrepancies between predicted and experimental values be accounted for?

In responding to these questions, it is appropriate for you to comment on the practical problems of data collection and instrumentation and to indicate how such problems might impact the overall project results. In addition to these questions, however, your instructors will usually require you to comment or reflect on particular questions about your projects, and you will often find that you can best respond to these requirements in a Discussion section that follows the results presentation.

Conclusions

The **Conclusions** section should summarize any important findings, results, or conclusions that you presented earlier in the report. You should not introduce any new results or interpretation in the Conclusions section. However, you may present any broader takeaways from the analysis and discussion in the report.

Figure 1 displays the first results of the fluid flow measurement project. Using a Laser Doppler Velocimeter, the fluid flow velocity inside a glass pipe was obtained at 16 locations across the two-inch diameter of the pipe. Measurements were obtained for laminar flow, with Reynolds number value of Re=380 and for turbulent flow, with Re=22600. The scale for turbulent intensity is represented on the right edge of the plot area. The laminar flow measurements are represented by diamond markers connected by a solid line, and turbulent measurements are represented by circular markers connected by a dashed line. Models for laminar and turbulent flow were prepared using Equation 1

$$u/U_c = 1 - \left[\frac{r}{R}\right]^2 \qquad (1)$$

where u/Uc is the normalized intensity, and r/R is the normalized radius of the pipe. The R value was obtained from measurements of the pipe, and the r values were read from the LDV's positioning gage after calibration. For each data set, values for u were obtained directly from the LDV measurements of flow velocity, and U values were compiled from the complete run of measurements. As the plot shows, the measured results for laminar flow compare well with the model. In contrast, the measurements of turbulent flow differ from the turbulent model's prediction and require further investigation.

Equation is cited

Inputs are defined

Conclusion is stated

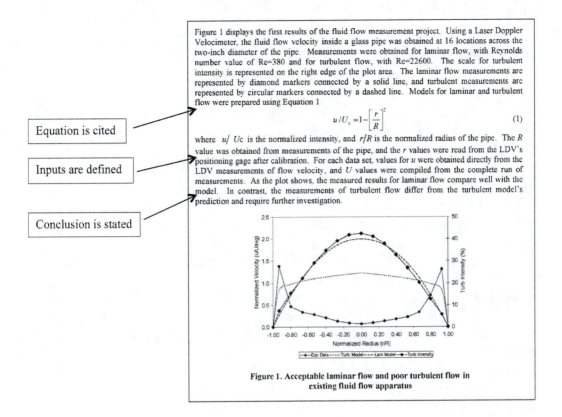

Figure 1. Acceptable laminar flow and poor turbulent flow in existing fluid flow apparatus

Report Format
A Quick Reference Sheet

Main headings are in bold. Underpinning questions are in roman type.

Introduction (sometimes called Objectives or Background)

    1. What was the overall goal of the project?

    2. What values or relationships were to be determined?

Apparatus and Procedures

    3. What instruments were used to collect data?

    4. What was the method for collecting data?

Results (also called Data and Findings)

    5. What was done to collect data? (Only for details not described under Procedures)

    6. What result, or data was obtained?

Discussion

    7. How do the results compare with expected or projected values?

    8. How are discrepancies between real and predicted values explained?

Conclusions

    9. What important findings are mentioned in the Results or Discussion section? What are the broader takeaways?

Sample Calculations

(not always a separate section)

Format and information in common report types

There are three general kinds of reports and supporting documents:

- Itemized Reports
- Narrative Reports
- Abstracts/Summaries

These report forms differ mainly in size, with Itemized Reports being short documents that describe very narrow projects, and with Narrative Reports being generally longer documents that describe large, open-ended projects. Abstracts and Summaries are usually short documents, usually prepared as attachments to longer reports. In the next sections of this book, we will provide details about format and information in each of these report types.

STYLE IN PARAGRAPHS AND SENTENCES

People use the term *style* loosely to describe the way you assemble words into sentences and paragraphs. In technical and scientific writing, we extend this to describe also the way that you prepare and describe visual displays. When people read your reports, they use your words and displays to learn about your project. They will best understand your work if you know how they read and what they expect. In this section we will briefly explain these things.

Specifically, in this section we explain how to write in a style that is scientifically appropriate and that is clear. Then we explain the norms for displaying data in plots, tables and simple calculations.

Scientific writing style

Instructors often confuse students by specifying the grammatical features of lab reports. Specifically, students are sometimes told that reports must be prepared in third person and in passive voice. Some instructors also specify the tense to be used in student reports, although instructors are split on the question of whether that tense should be present or past. While such grammatical guidance speaks to real concerns in professional communication, the issues are complex and call for a more nuanced approach to instruction and grading.

In scientific reports you use words to describe reality, and you must be flexible in your word selection because the events and phenomena you observe will be complex. You should expect to select words from different grammatical categories to describe the different parts of your projects and the different phenomena that you observe. You will select verbs in order to describe the actions that took place during your project, and you will set the tenses for those verbs in order to accurately represent when those actions took place. You will select a grammatical person (first person or third person) for each of those verbs. You will need to vary your selection of person, verbs, and tenses in order to reflect the various events that take place during a project and the changing status of information that you are presenting to the reader.

Tense. Tense is a good example of this grammatical variation. In most cases, undergraduate project reports describe work that has been completed, so most undergraduate reports should be prepared using *past tense*. However, it is perfectly acceptable to change tenses so long as the change introduces information that is logically different from the past-tense project description. For example, figures are conventionally described using *present tense* verbs, because readers read the text description at the same moment as they view the figures. In a similar way, mathematical proofs are usually presented in present tense text, as readers are expected to view the proof text as a set of instructions to be followed as they read.

Other acceptable tense shifts may describe background information—events that began in the past and are ongoing—or they may describe universal truths that can be derived from your data. If, for example, your report addresses a problem that many people have worked on previously, you might describe your work in this way:

Premature failures due to fatigue <u>have been</u> reported in installed Modular Bridge Expansion Joints (MBEJs) [2 & 3]. These failures require either repair or replacement. Exhaustive research by Dexter et al. [4] <u>examined</u> the behaviour of MBEJs under static and cyclic loading. Among many issues needing more research, as stated in NCHRP report 467, is a fatigue-resistant field splice for centre-beams. To minimize traffic blockage, the replacement of failed MBEJs needs to be done <u>in several stages</u>, during which one or two lanes may be closed. The installation of an MBEJ will typically be staged and therefore need to be spliced in the field. It is not always possible to have optimum quality control over the splice installation due to the limited working space and time constraints (Fig. 2). The current designs for centre-beam field splices, whether of the hinge-type (Fig. 3) or moment-resisting type (Fig. 4), do not possess the same fatigue life as that of the other components of the MBEJ and thus cause premature, costly failures in the installed MBEJs. Chaallal et al., thus, recommend "an improved splice be designed" (2002) [5].

[from MacDougall, example report in Part 3 of this book]

Here, past tense verbs have been underlined, present tense verbs are in boxes, and future tense verbs are circled. This blend of past and present tenses is justified by logic. When you change tenses in your project reports, you simply need to be sure that there is a real and logical reason for that change. Thus, when you describe a project that is still in progress, you should use present tense verbs to describe work that is ongoing as well as past tense verbs to describe tasks that have been completed. When you prepare a proposal, you will naturally use future tense as well as past and present tense to distinguish work that has not been started from work that may be complete or ongoing.

Person.

As indicated above, mathematical proofs are conventionally presented in present tense; they are also presented in second person. This means that no matter what your instructor tells you to do, you cannot fully present analytical work in third person. In fact, it is better to say that your instructors want your scientific reports to use *impersonal* style. This means that you should avoid speaking of yourself and your teammates when you describe your project work.

Grammatical person in a sentence describes who or what is the subject of the sentence and is the agent of the verb's action. While you as a scientist may take many actions in designing an experiment, your task in a report is to describe the behavior of the objects you are studying and the performance of the instruments you used. Because your reports should describe nature, not you, the subjects of your sentences should be the objects and forces you are studying and the instruments you use to study those objects and forces.

The following sample paragraph shows how a team incorrectly uses first-person sentences to describe their project work:

> *The team* was asked to determine the RMS and peak-to-peak voltages. *We* were uncertain at first how to measure these voltages, but after discussion with *our teaching assistant, we decided* to measure these voltages directly from the oscilloscope screen. *We decided* to conduct this part of the experiment with the frequency set to 1 kHz on the signal generator, and *we decided* to set the CD component to 3 volts. After many adjustments and false starts, *we measured* a peak voltage of 0.84 volts. *Then we* found the RMS value to be 0.29 volt.

In this example, the first-person subjects have been italicized. This paragraph is inappropriate as a project description. The sentence subjects call attention to the team that performed the experiment; in order to describe the project information while using these first person sentences, you must write needlessly long sentences and paragraphs. At best, the first person statements waste space and time; under some circumstances, such needless information can actually disrupt the presentation of actual project results.

Below we revised this paragraph to create an appropriately impersonal paragraph. The verb forms characteristic of this impersonal stance are set in italics:

> The RMS and peak-to-peak voltages *were directly measured* on the oscilloscope, using a frequency of 1kHz on the signal generator, with the DC component of the signal set to 3 volts and the AC component oscillating at 1kHz. The peak-to-peak value *was found to be* 0.84 volts and the RMS value *was found to be* 0.29 volt.

Clearly, this impersonal presentation is shorter than the original version, and this is an important advantage for readers of scientific reports.

People appear in scientific reports only rarely and in specific circumstances. In research reports these circumstances usually involve a background statement, which may describe projects that other scientists have undertaken and that are relevant to the current research subject. These project descriptions usually identify the primary researchers by name in a single sentence; after this initial reference to a scientist, the sentence subjects in the project descriptions revert to terms that name the experiment's instruments and results.

Passive voice.

Passive voice is a sentence form that scientists use to make their reports impersonal, because passive constructions enable authors to drop people from their sentences. The sentence below, from the example first-person paragraph above, uses an active verb (underlined) with a first-person subject (italicized) and a simple object (**bolded**):

> *Then we* <u>found</u> **the RMS value** to be 0.29 volt.

To make this sentence impersonal, scientists typically begin by creating a passive verb, as in this sentence:

> Then *the RMS value* <u>was found</u> **by us** to be 0.29 volt.

When you turn a verb into a passive, it is necessary to invert the sentence. The newly passive verb is <u>underlined</u> in the second example. In addition, this verb has moved to the right in the sentence; the object of the first sentence, *RMS value,* has become the subject of the new sentence; it is italicized. Finally, the subject of sentence 1, *we*, has become an object, **us**, in **bold**.

Clearly, you use passive constructions to rearrange the words in sentences. Now, when the main agent has become an indirect object in the sentence, our grammar allows us to remove that structure from the sentence entirely. The resulting passive sentence is shown as in the third version of our sentence:

> The *RMS value* <u>*was found*</u> *to be* 0.29 volt.

In this sentence the first person subject has been removed, to be replaced by a subject that describes an experimental result. This sentence is an impersonal sentence; it is short, it is professionally direct, and it is project-focused.

While passive constructions are common in scientific writing, it is possible to write impersonal reports with active verbs. Some examples follow:
1) Figure 1 *demonstrates* that the model accurately predicts the system response.
2) The DC offset dial *controls* the magnitude of the average, or mean, voltage.
3) The experimentally measured values *are* higher than the predicted values.
4) The calculated values of water jet deflection *correspond* to motion of the jet's centerline.

The verbs are italicized in each of these example sentences. These sentences are impersonal—the subjects are experimental tools and results rather than people—and they also take active verbs.

The lesson to learn from these examples is that you can reasonably use both active and passive verbs in your technical or scientific reports. The verb's grammatical form is not the primary concern for readers. Rather, readers want you to describe the experiment in objective, impersonal terms. When you do this, your reports will not only sound professional, they will also be short, and they will be focused on the substance of your project.

CHAPTER **1.2**

ABSTRACTS, ITEMIZED REPORTS AND NARRATIVE REPORTS

There are many types of scientific reports. In this book we simplify the field of report types to focus on the three most common and robust types of reports that technical professionals produce:

- **Abstracts** (called **Summaries** in some circumstances)
- **Itemized Reports**
- **Narrative Reports**

These report types, or templates, differ in size, with narrative reports tending to be long documents, and with abstracts and itemized reports being sharply limited in size. However, these size differences are driven by differences in the ways these documents are used and the types of projects for which they are considered appropriate.

Regardless of their size, your technical documents will be read by people who need to see your answers to these questions:

1) What was the goal of the work?
2) How was data collected and analyzed?
3) What results were obtained?
4) What are the possible sources of error?
5) What is the significance of the work?

These questions roughly correspond to the main headed sections of a report, as described in the last chapter. While instructors and supervisors may ask you to use any of several different document templates for your reports—such as the itemized, letter and memorandum formats that we present—these different templates are simply used to package your information for particular audiences. Your documents will always provide the same kinds of information; you will simply present your work in more or less detail according to the document template you use. Here we will briefly review how these different document templates are used and how you can prepare your documents to best address their expected use.

Abstracts and Summaries

Abstracts and Summaries are short documents that usually condense the points of lengthy narrative reports. These are usually attached to the front of the reports they describe in order to provide readers a concise index to the information contained in the body of the report. These documents are convenient for readers who wish to understand and discuss a project whose full report may be too large to reproduce or read in a timely way. Such readers may use an Abstract or a Summary to make decisions that do not require detailed understanding of the report, or they may use it as a memory aid after they have finished reading the complete report.

Abstracts and Summaries are often described as "stand-alone" documents. This means that readers expect these to completely account for the essential points of your work. The project's background, goals and methods should be compressed into a sentence or two, and the project's critical results should be presented and explained, as should any possible sources of error and any open questions. When appropriate, the impact of the work—on future research or on business scheduling—may be defined.

Abstracts and Summaries must be short. Abstracts are commonly restricted to 250 words, while Summaries are frequently restricted to a single printed page. Often they are presented as a single paragraph. These size restrictions can be adjusted for very large projects, but this is seldom justified for undergraduate laboratory reports. Because space for abstracts and summaries is sharply limited, tables, plots and references should be avoided in these documents. Critical results should be presented in numerical form with minimal discussion; readers who desire details concerning particular results are expected to read the full report.

Abstracts and Summaries address different types of audiences and contexts. Abstracts are mainly attached to reports that emphasize results alone. Summaries are usually attached to longer narrative reports, where background and future work must be addressed as well as the results of the project work.

In Abstracts for your lab reports, you should make four kinds of statements:
1) The **Goal** of the work
2) The **Methods** of data collection
3) The **Results** that were obtained (both measured and analytical results)
4) The **Conclusions** (or takeaway points)

The sample abstract on the following page was prepared for a capstone project report, and it demonstrates all of the features required in an appropriate report abstract:

A Summary is a long abstract. A Summary differs from an abstract in that a summary provides a more detailed problem statement (often with background or motivation), it may provide a more detailed description of the methods, and it may conclude with action-item recommendations, including suggestions for future research, estimates of budget, and even production schedules. Summaries may also present the report's results with some detailed discussion. Consequently, Summaries are commonly viewed as stand-alone documents, or as very small reports.

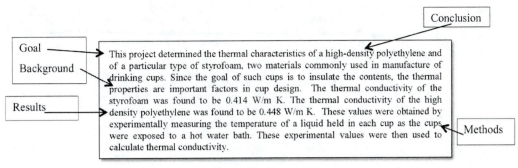

This project determined the thermal characteristics of a high-density polyethylene and of a particular type of styrofoam, two materials commonly used in manufacture of drinking cups. Since the goal of such cups is to insulate the contents, the thermal properties are important factors in cup design. The thermal conductivity of the styrofoam was found to be 0.414 W/m K. The thermal conductivity of the high density polyethylene was found to be 0.448 W/m K. These values were obtained by experimentally measuring the temperature of a liquid held in each cup as the cups were exposed to a hot water bath. These experimental values were then used to calculate thermal conductivity.

Example Abstract

In your undergraduate classes, most report Summaries will be between three-quarters of a page and one full page in length. A one-page Summary may be divided into three or four paragraphs, with each paragraph presenting one of the fundamental statements of a project report, as follows:

1) The Goal and/or background of the work
2) The Methods of data collection
3) The Measured and Analytical Results that were obtained
4) Conclusions, significance or takeaway points

The sample Summary on page 20 is drawn from one of the example reports used in this book; it describes an investigation performed in an intermediate experimentation course.

Preparing itemized reports

Professionals prepare Itemized reports to describe projects that are narrowly defined and that have a specific and short list of tasks. In industry, for example, itemized reports are used to describe procedures that are both repetitive and common, such as qualification or acceptance tests. These tests may be conducted according to an established professional code or protocol that defines procedures, methods and acceptable result values. When well-established guidelines govern a project, it is often unnecessary to rehearse them in the project report; the itemized report will focus on the measured results, while providing small spaces for disclosure of non-standard conditions or procedures. For projects of this sort, companies may even rely on pre-printed itemized report forms or worksheets.

In undergraduate experimental projects, you may use the Itemized report format to describe pre-lab homework projects as well those introductory projects whose procedures have been pre-defined for you. In itemized reports, you will be asked to briefly describe the goal, apparatus and procedures, but the great majority of your report space will be devoted to presenting and describing your results.

In your undergraduate itemized project reports, you will often be expected to show your work—motivating the project tasks, and describing and discussing the results—in order to

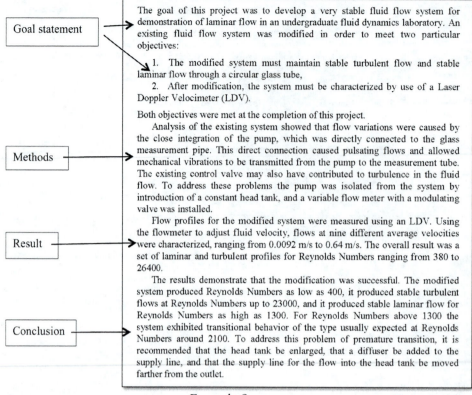

The goal of this project was to develop a very stable fluid flow system for demonstration of laminar flow in an undergraduate fluid dynamics laboratory. An existing fluid flow system was modified in order to meet two particular objectives:

1. The modified system must maintain stable turbulent flow and stable laminar flow through a circular glass tube,

2. After modification, the system must be characterized by use of a Laser Doppler Velocimeter (LDV).

Both objectives were met at the completion of this project.

Analysis of the existing system showed that flow variations were caused by the close integration of the pump, which was directly connected to the glass measurement pipe. This direct connection caused pulsating flows and allowed mechanical vibrations to be transmitted from the pump to the measurement tube. The existing control valve may also have contributed to turbulence in the fluid flow. To address these problems the pump was isolated from the system by introduction of a constant head tank, and a variable flow meter with a modulating valve was installed.

Flow profiles for the modified system were measured using an LDV. Using the flowmeter to adjust fluid velocity, flows at nine different average velocities were characterized, ranging from 0.0092 m/s to 0.64 m/s. The overall result was a set of laminar and turbulent profiles for Reynolds Numbers ranging from 380 to 26400.

The results demonstrate that the modification was successful. The modified system produced Reynolds Numbers as low as 400, it produced stable turbulent flows at Reynolds Numbers up to 23000, and it produced stable laminar flow for Reynolds Numbers as high as 1300. For Reynolds Numbers above 1300 the system exhibited transitional behavior of the type usually expected at Reynolds Numbers around 2100. To address this problem of premature transition, it is recommended that the head tank be enlarged, that a diffuser be added to the supply line, and that the supply line for the flow into the head tank be moved farther from the outlet.

Example Summary

display your understanding of the project. To that end, you should include the following headed sections in your itemized reports.

1. **Objectives**
2. **Apparatus** and **Procedures**
3. **Results** (sometimes called **Findings**)
4. **Analysis** (often integrated with the Results presentation)
5. **Discussion**
6. **Conclusions**

Sample itemized reports from an introductory experimentation course are included in sections 1.3 and 1.4 of this book. You will see that these examples skew heavily towards the reporting of results, with very brief descriptions of the goals, apparatus and procedures.

We will now turn our attention to a model of a complete lab assignment from an introductory experimentation course. The project is titled "Lab Zero: Introduction to the Oscilloscope," and the model presents the specific project requirements in the form of a lab instruction sheet, shown at the beginning of Chapter 1.3. The model report then describes the action taken on these requirements. To better focus on the details of text and display format, the cover sheet is omitted in this example. Chapter 1.4 then presents a second example of a short itemized report.

CHAPTER 1.3

LAB MANUAL FOR OSCILLOSCOPE INVESTIGATION

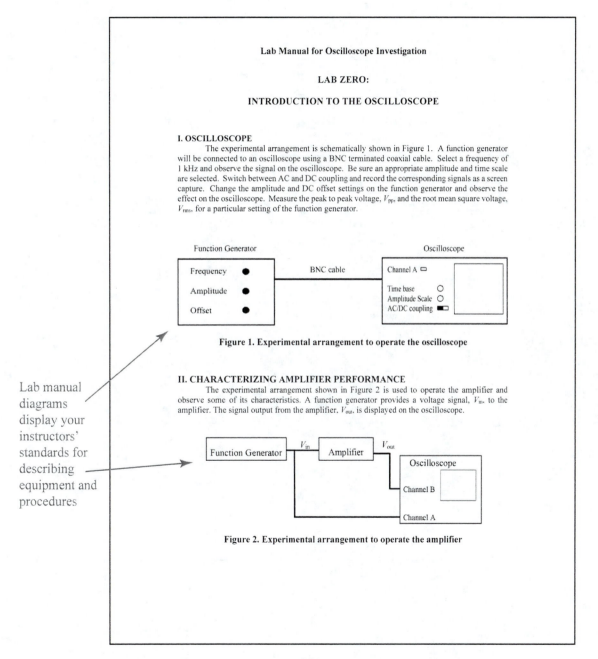

Lab Manual for Oscilloscope Investigation

LAB ZERO:

INTRODUCTION TO THE OSCILLOSCOPE

I. OSCILLOSCOPE

The experimental arrangement is schematically shown in Figure 1. A function generator will be connected to an oscilloscope using a BNC terminated coaxial cable. Select a frequency of 1 kHz and observe the signal on the oscilloscope. Be sure an appropriate amplitude and time scale are selected. Switch between AC and DC coupling and record the corresponding signals as a screen capture. Change the amplitude and DC offset settings on the function generator and observe the effect on the oscilloscope. Measure the peak to peak voltage, V_{pp}, and the root mean square voltage, V_{rms}, for a particular setting of the function generator.

Figure 1. Experimental arrangement to operate the oscilloscope

II. CHARACTERIZING AMPLIFIER PERFORMANCE

The experimental arrangement shown in Figure 2 is used to operate the amplifier and observe some of its characteristics. A function generator provides a voltage signal, V_{in}, to the amplifier. The signal output from the amplifier, V_{out}, is displayed on the oscilloscope.

Figure 2. Experimental arrangement to operate the amplifier

Lab manual diagrams display your instructors' standards for describing equipment and procedures

Data Collection
Sheet: Students
may manually
record data here,
for later transfer
to notebooks

Lab Manual for Oscilloscope Investigation

A. DC operation (*i.e.*, when V_{in} and V_{out} are constant) .

Record the output voltage amplitude of the amplified signal, V_{out}, when sweeping the input voltage between −12 and + 12 volts.

V_{in} (volts)					
V_{out} (volts)					

B. AC operations, *i.e.*, when the input voltage is oscillating at a given frequency.

Set the function generator to "sine wave" with zero offset. Measure the frequency response of the amplifier between 10 Hz and 1 MHz on a log scale, *i.e.*, at 10, 100, 10^3, 10^4, 10^5, and 10^6 Hz by measuring the ratio $K = (V_{out} / V_{in})$, where V_{out} and V_{in} are peak-to-peak voltages. K is the amplification factor and $G = 20 \log_{10} K$ is defined as the gain in decibels. Note that K and G may vary with frequency.

C. Clipping

Set the frequency to 1 kHz. Increase the input voltage and record the resulting waveform, *i.e.*, the shape of the amplified signal, as a screenshot.

Record your input values, measured output values and calculated values using the table below.

Frequency (Hz)	V_{in} (volts)	V_{out} (volts)	K	G (dB)

Companies may
provide
standardized forms
such as this for
data collection on
repetitive tasks,
such as quality
assurance tests

SAMPLE LONG ITEMIZED REPORT

I. DYNAMIC ELECTRICAL MEASUREMENTS

A. Objectives and Procedures

The objectives of this experiment were to observe the effect of a direct current (DC) offset on an input signal, and to describe the effects of alternating current (AC) and DC coupling on the output of an oscilloscope.

To reveal the functions of the DC offset, AC coupling, and DC coupling, the voltage generated by a Gwinstek SFG-2100 function generator was observed directly on a Tektronix TPS 2012B digital oscilloscope. Screen capture images were made of waveforms to record the effects of these functions on the output signal. Subsequently, using a department-fabricated, non-inverting, constant gain amplifier, mean voltage, peak-to-peak voltage, RMS voltage, and frequency were measured with the oscilloscope to demonstrate how changes in these settings can impact measurements of output signals.

B. Experimental Results

The DC offset control on the function generator was shown to control the magnitude of the average, or mean, voltage. It was also noted that neither the peak-to-peak voltage nor the frequency varied with DC offset. The RMS voltage did vary, but this is expected, since it is related to the average absolute value of the voltage. Table 1 displays the measured quantities for a sinusoidal signal without DC offset and a sinusoidal signal with a 3-volt DC offset.

Table 1. Measured values for sinusoidal signals with and without DC offset

	With offset	Without offset
DC offset (V)	0.00	3.00
Mean voltage (V)	0.00	3.00
Peak-to-peak voltage (V)	0.84	0.84
Frequency (kHz)	1.00	1.00
RMS voltage (V)	0.29	3.02

1

Experimental equipment is specified as fully as possible

Captions are bolded to distinguish them from surrounding text

Citations define:
1) Which display readers should look at
    2) What information the display presents

Figure 1 is a screen-capture image of a sine wave signal with zero DC offset, whereas Figure 2 is a screen-capture image of the same signal with a 3-volt DC offset. Note that the vertical scales in the two figures are different.

Plots are large in this sample report to enable easy grading by instructors

Clients and industrial supervisors may prefer smaller displays

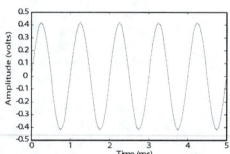

Figure 1. Sinusoidal (AC) voltage with no constant (DC) offset

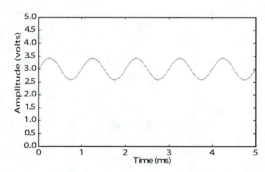

Figure 2. Sinusoidal (AC) voltage signal after a 3-volt DC offset has been added

2

To characterize the operation of an amplifier, a non-inverting, constant-gain amplifier was added to the system, with its output directed to the oscilloscope's Channel B. The signal from Figure 2 was transmitted from the function generator, and the oscilloscope was set to DC coupling. Figure 3 is a screen capture of the waveform observed on the oscilloscope with DC coupling activated.

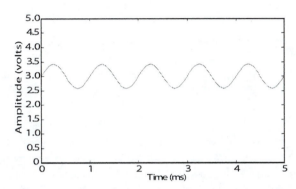

Figure 3. Waveform obtained with DC coupling, showing sinusoidal (AC) voltage offset by a constant (DC) voltage

To observe AC coupling, the input signal was held constant, but the oscilloscope setting was changed from DC to AC coupling. While the input signal still had a 3 volt DC offset, the waveform on the oscilloscope screen was centered on 0 volts. Figure 4, below, is a screen-capture image of the waveform observed on the oscilloscope when AC coupling was used.

When a display is large, it may force an early page break, leaving a large blank space, as at the bottom of this page. You should avoid such large blanks when possible. Consult with your instructor or supervisor to determine whether such blank spaces are of concern and how you are to manage such problems of page layout.

3

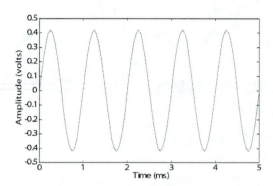

Figure 4. Waveform obtained with AC coupling, showing no apparent offset

Figures 1 and 2 demonstrate that the DC offset control on the function generator controls the magnitude of the mean voltage of the signal. When there was no DC offset, the average value of the signal was zero volts. Because the positive magnitudes and the negative magnitudes of the signal were identical values, it is possible to say that the signal was "centered on" zero volts. By adding a DC offset of 3 volts, the signal was "moved up" to be "centered on" 3 volts.

Figure 3 shows that a signal measured with DC coupling displays both the DC component and the AC component of an input signal. In contrast, Figure 4 indicates that a signal measured with AC coupling shows only the AC component of the input signal. When AC coupling is selected, the DC offset is eliminated by the use of a capacitor or some other arrangement of circuitry. This setting may create distortions in measurements of very low frequency oscillations, as the output signals may be treated as approximately constant, or DC voltage, depending on the time constant of the circuit.

C. Discussion and Analysis

In this experiment, voltage was obtained using two different methods: Peak-to-peak and root mean square (or RMS). The peak-to-peak voltage is found as the difference between the maximum and minimum values of an oscillating signal. The root-mean-square voltage is obtained through a

4

Reducing the size of this display could eliminate the blank space on the previous page more complicated process. Its name is derived from the mathematical calculation used to determine the RMS value.

Formal Equation Presentation:

In order to use the measured values to obtain the RMS value for voltage, first let the sinusoidal input signal with no DC offset, $V(t)$, be described by Equation 1,

1) Citation

$$V(t) = A\sin(\omega t) \qquad\qquad 1$$

2) Symbolic Display

where A is the signal amplitude, ω is the angular frequency, and t is time. The root-mean-square of the signal, V_{rms}, is defined by Equation 2,

3) Definition of Variables

$$V_{rms} = \left[\frac{1}{T}\int_0^T V(t)^2 \, dt\right]^{1/2} \qquad\qquad 2$$

where T is the period of the signal. For a specific time interval, the signal amplitude is squared at each moment. The mean of these squared values is evaluated using the integral. The square root of this mean is the RMS value.

You need not re-define variables that have been defined earlier in your report

Substituting Equation 1 into Equation 2 gives the relationship between the amplitude of a sinusoidal signal and its RMS value:

$$V_{rms} = \left[\frac{1}{T}\int_0^T V(t)^2 \, dt\right]^{1/2} = \left[\frac{1}{T}\int_0^T A^2\sin^2(\omega t) \, dt\right]^{1/2} = \frac{A}{\sqrt{2}} \qquad\qquad 3$$

The measured peak-to-peak voltage was 0.84 volts. The amplitude, A, is half the peak-to-peak voltage, V_{pp}, and

$$V_{rms} = \frac{V_{pp}}{2\sqrt{2}} = \frac{0.84 \text{ volt}}{2\sqrt{2}} = 0.297 \text{ volt} \qquad\qquad 4$$

In advanced classes, it is not necessary to show the substitution of measured values

The measured RMS value of 0.29 volts is within 3% of the above-calculated value. The difference between measured and predicted values could be due to inconsistencies in the input signal. However, no fluctuations in the measured RMS value were noted during the experiment.

5

In the Conclusion section, you should restate the points made for each project task, and provide a takeaway point or conclusion

D. Conclusion

Using a function generator with a 1 kHz AC output signal, the oscilloscope's 3-volt DC offset was observed, and its display was compared to the display of a 1kHz signal with no DC offset. Neither the peak-to-peak voltage nor the frequency varied with the DC offset. An amplifier was introduced to the system, and the RMS voltage was found to vary with the DC offset. This variation was expected, however, as RMS voltage is related to the absolute value of the voltage.

6

CHAPTER 1.4

EXAMPLE OF A SHORT ITEMIZED REPORT

Memorandum format defines the display of Supervisor, Author(s), Title and Submission Date

The Short, Itemized report is useful for presenting results that require no motivation and no discussion. Pre-lab assignments are often submitted as short, itemized reports.

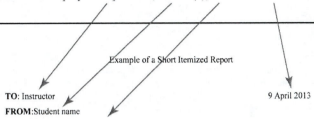

Example of a Short Itemized Report

TO: Instructor   9 April 2013

FROM:Student name

SUBJECT: Report for Calorimeter Experiment

Experimental Spreadsheet. An EXCEL spreadsheet was prepared to receive and process the experimental data. The spreadsheet, including representative data, is appended as Attachment 1. A block of the pertinent data is reproduced below as Table 1.

Table 1. Heat capacity data and model

temp	data	model
C	J/kg-K	J/kg-K
10.1	2005	2000
20.0	2030	2041
29.9	2105	2100
40.2	2180	2182
50.0	2285	2278
59.9	2390	2394

Regression Modeling. A regression model of the form, was developed for the representative data. It is presented as Equation 1:

$$C_p = \left(\frac{dH}{dT}\right)_P = C_0 + B_1\,T + B_2\,T^2 \qquad\qquad 1$$

The model and the data are plotted in Figure 1.

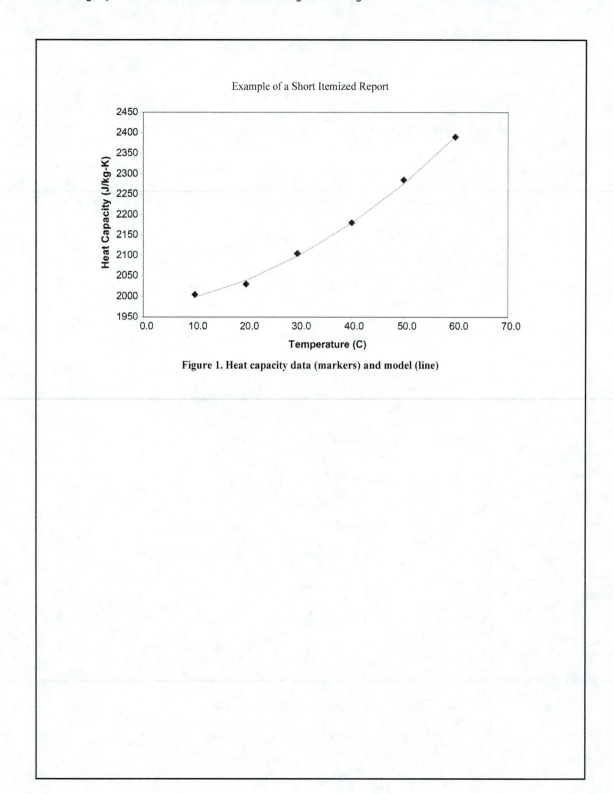

Example of a Short Itemized Report

Figure 1. Heat capacity data (markers) and model (line)

Example of a Short Itemized Report

Attachment 1. Example experimental spreadsheet

File: HEATCAP		29 Sep 2013, SMJ				
Example spreadsheet illustrating data processing, regression, and presentation.						
Summary Results:						
	alpha risk =	0.007				
	R squared =	0.998				
Example Data:					1	2
	temp	data	model		tc	tc^2
	C	J/kg-K	J/kg-K			
	10.1	2005	2000		10.1	102.0
	20.0	2030	2041		20.0	400.0
	29.9	2105	2100		29.9	894.0
	40.2	2180	2182		40.2	1616.0
	50.0	2285	2278		50.0	2500.0
	59.9	2390	2394		59.9	3588.0

SUMMARY OUTPUT						
Regression Statistics						
Multiple R	0.998976397					
R Square	0.997953841					
Adjusted R Square	0.996589735					
Standard Error	8.766268858					
Observations	6					

ANOVA						
	df	*SS*	*MS*	*F*	*Significance F*	
Regression	2	112440.2909	56220.14546	731.5809576	9.25569E-05	
Residual	3	230.5424091	76.84746969			
Total	5	112670.8333				

	Coefficients	*Standard Error*	*t Stat*	*P-value*	*Lower 95%*	*Upper 95%*
Intercept	1976.921714	15.82745332	124.9045992	1.13145E-06	1926.551646	2027.291781
X Variable 1	1.303830113	1.03686987	1.257467451	0.297565826	-1.99595567	4.603615896
X Variable 2	0.09445367	0.014508043	6.510434915	0.007360909	0.048282559	0.140624782

CHAPTER 1.5

NARRATIVE REPORTS

Itemized reports are prepared to describe projects that are repetitive and narrowly defined. In contrast, Narrative reports are prepared to describe projects that may be large and whose solutions may not be known in advance. Consequently, a narrative report must be designed to tell the whole story of a project, beginning with the motivation for the work and concluding with a statement of the significance of the conclusions that were obtained.

A narrative report should provide a complete account of your work on a project, with the information subdivided into headed sections as described in Chapter 1.1:

1. Introduction, to describe the goals, related background, and scientific approach
2. Methods, to describe the apparatus, procedures and any benchmarks or models that may be used for comparison purposes
3. Results, to describe measured and analytical results
4. Discussion, to evaluate the quality of the results in comparison with benchmarks
5. Conclusions, to summarize and review the points presented in the Results and Discussion sections

In classrooms, this information is typically packaged in a formal report that uses a cover sheet and Table of Contents, as does the example report in Chapter 1.9. However, two variant forms, the Letter Report and the Memorandum Report, are often seen. These variant report forms differ mainly in the ways that the author and supervisor are displayed at the head of the report.

In letter reports, as presented in Chapter 1.7, the first page of the report is formatted as a business letter, with the author(s) identified in the letterhead or return address block, and the recipient identified in the addressee block of the letter. The Introduction of the report begins immediately following the letter's salutation statement. The final page of the letter report is expected to provide a closing and signature block, as does a business letter. The letter report format is most commonly used in professional correspondence between consulting engineers and their clients.

In memorandum reports, as presented in Chapter 1.5, the first page of the report is formatted as a formal business memorandum, with **TO**, **FROM**, **SUBJECT** and **DATE** blocks positioned just above the report's Introduction statement. Memorandum reports are typically used in professional correspondence between engineers and their supervisors. In these situations, the engineer authors' names are listed beside the FROM heading, the supervisor names are listed beside the TO heading, and report's title is placed on the SUBJECT line.

The differences between letter reports, memorandum reports and formal reports are mainly found in the layout of a document's first page. Using editing software, these format matters can be easily managed, and they may appear to be formalities. However, these formalities concern important information. In filling the authorship, readership, date and title slots at the beginning of a report, you define the lines of responsibility and points of contact. These definitions must be accurate, as they will be carefully examined in the event of a dispute between authors and clients or supervisors.

Narrative reports can vary in size, according to the scope of the project they may describe. Short reports typically describe small projects (or project tasks) that were developed under supervision. If a student happens to work on a faculty member's research project, for example, that student might be given a general goal or a question to answer, and then be left to design and conduct an experiment that meets that goal or answers that question. The short report form is used to describe projects in an intermediate experimentation course.

People prepare long reports to describe large or open-ended projects that have been conducted independently. In such situations, the author of the report is usually the principle investigator on the project, and the report's readers generally are not completely familiar with the project's technical details. Consequently, long reports are common on research projects, even if a supervising faculty member originally assigned the project. As research supervisors, faculty members often find it easier to read complete reports, and from time to time they pass these reports along to colleagues who are less familiar with the particulars of the project. A project report is required as the final submission in most capstone courses, as it presents the results of a student team working over the course of an entire term.

Long reports are appropriate under either of two conditions. First, long reports are appropriate for large and complex projects, for the author may have many procedures to describe, and the author must also explain how all the parts of the project fit together. Second, long reports are necessary for most readers who have not been involved in the daily work of the project. If a student completes a project during one afternoon in a well-supervised lab, then that student probably will not write a long report. If, however, the project was spread across most of a term, then a long report is probably expected.

We will now turn our attention to sections 1.6, 1.7, 1.8 and 1.9, which present long report examples.

CHAPTER 1.6

EXAMPLE OF A SHORT NARRATIVE
REPORT IN MEMORANDUM FORMAT

The Standard School of Mechanical Engineering
Genuine Institute of Technology, College Town USA 55511-0001
Undergraduate Instructional Laboratories

TO: Undergraduate Lab Instructional Staff   31 January 2013

FROM: Ernest L. Scholar, Section A

SUBJECT: Experiment on the Period of a Pendulum

Background and Objective

The objective of this experiment was to experimentally measure the period of a simple pendulum and to compare the obtained data with an estimate of the period based on calculations. The experiment was conducted in the instructional lab on 8 January 2013.

Apparatus and Procedure

Apparatus

The pendulum is a locally-fabricated flat steel bar with a drilled hole for the pivot, which is a knife edged shape. The pendulum used for these measurements is identified by Serial No. 3. The dimensions of the pendulum were measured with a Starrett model No.123 vernier caliper, with an uncertainty of .05 mm (.002 inch), and with a True Value MMS425 measuring tape, with an uncertainty of 1 mm (1/32 inch). The period was measured with a Timex Indiglo quartz crystal wrist chronometer. The fractional uncertainty of the chronometer is 6.0×10^{-6} (15 seconds/month [1]).

Procedure

The period of the pendulum was estimated by using the measured values for its dimensions and mass. To experimentally measure the period of the pendulum, the pendulum was placed on a pivot and set in motion. The period was measured by manual timing for four repeated 50-cycle tests. The results and analysis of these tests are given below.

Measurements

Dimensional Measurements

The obtained dimensions of the pendulum were as follows. Using the measuring tape, the overall length was found to be 349 mm. Using the vernier caliper, the thickness was measured to be 13.1 mm, and the width was measured to be 48.86 mm. Also using the vernier caliper, the diameter of the hole was measured to be 19.9 mm, and the distance from the edge of the hole to

1

Subheads for individual paragraphs are common in practice but are not required for class projects

This is a reference citation using IEEE format

Instruments can be efficiently specified at the time a measurement is introduced

34

Table numbers and captions are placed above tables

Table entries are best aligned on the decimal point

the end of the bar was measured to be 10.0 mm. This pendulum is illustrated, with dimensional measurements, as Figure 1 in Attachment 1.

Measurements of the Period

The period was measured by timing 50 cycles with the wrist chronometer. The timing interval was defined to begin and end when the leading edge of the pendulum passed a small LED light source. The results of four period measurement tests are given in Table 1.

Table 1. Data for four measurements of fifty cycles each

test	time	period
	sec	sec
1	48.16	0.9632
2	48.13	0.9626
3	48.25	0.9650
4	47.76	0.9552
average =		0.9614
std dev =		0.0043
uncertainty of data =		0.014
uncertainty of average =		0.007

Results and Analysis

Using the dimensional data presented above, and using the formula presented by Anderson [2], the predicted period of the pendulum was calculated to be 0.9575 sec. Based on the uncertainties of the dimensional data as presented in Table 1, and using the method of Kline and McClintock [3], the uncertainty of the predicted period was found to be 0.0014 sec.

The average period, the standard deviation, and the uncertainty of the period data were computed and are shown in Table 1. The uncertainty of the data, U_{data}, due to random variation was computed using the relatively large coverage factor of 3.2 that applies in this case. The uncertainty of the average was obtained using the standard formula for error propagation analysis, as in Equation 1.

2

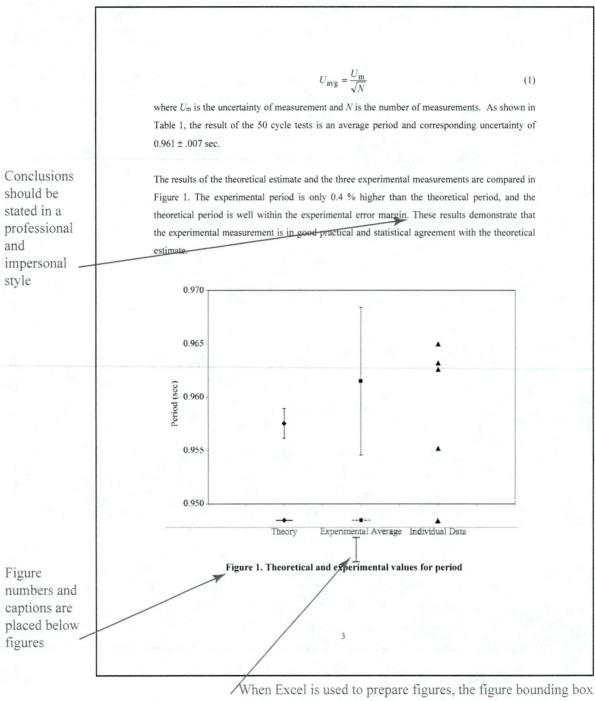

$$U_{avg} = \frac{U_m}{\sqrt{N}} \qquad (1)$$

where U_m is the uncertainty of measurement and N is the number of measurements. As shown in Table 1, the result of the 50 cycle tests is an average period and corresponding uncertainty of $0.961 \pm .007$ sec.

The results of the theoretical estimate and the three experimental measurements are compared in Figure 1. The experimental period is only 0.4 % higher than the theoretical period, and the theoretical period is well within the experimental error margin. These results demonstrate that the experimental measurement is in good practical and statistical agreement with the theoretical estimate.

Figure 1. Theoretical and experimental values for period

3

Conclusions should be stated in a professional and impersonal style

Figure numbers and captions are placed below figures

When Excel is used to prepare figures, the figure bounding box often creates an untidy blank space. Because this is created by the program, studnets should consult with their instructors to determine how to manage this blank space

36

In a short report, the Discussion and Conclusions sections may be combined.

Discussion and Conclusions

Despite the good general agreement between the measured results and the predictions, it is apparent that the experimental measurements tend to be higher than the predicted value. In fact, if the unaccountably low value of 0.9552 obtained on test 4 were excluded, the experimental average would be significantly higher statistically. The predicted values could be too low for at least two reasons: 1) there may be errors in the input dimensional data, or 2) the theory on which the prediction is based may be deficient or incomplete. The first of these reasons can be dismissed; the dimensional measurements are simple and are therefore taken to be reliable. Further, because the LED timer is very accurate, bias error in the experimental measurements is minimized. Consequently, discrepancies between the predicted results and the experimental results must be attributed to physical phenomena that are not accounted for in the simple model that was used to obtain the predicted values. For example, this theoretical model ignores friction, and it is possible that a small but non-zero value of friction could be found at the pivot. Additional experimental data would need to be taken to reduce the random uncertainty in the measurements and resolve this question.

This reference list uses the IEEE format style

References

[1]   T. Corporation. Instructions: Timex Quartz Digital Chronometer [Online]. Available: http://www.timex.com/instructions/086Instr.pdf

[2]   J. L. Anderson, "Approximations in Physics and the Simple Pendulum," *American Journal of Physics,* vol. 27, pp. 188-189, 1959.

[3]   S. J. Kline and F. McClintock, "Describing uncertainties in single-sample experiments," *Mechanical engineering,* vol. 75, pp. 3-8, 1953.

See Chapter 2.6 for more information about reference format and style guides

4

Attachment 1. Pendulum Diagram

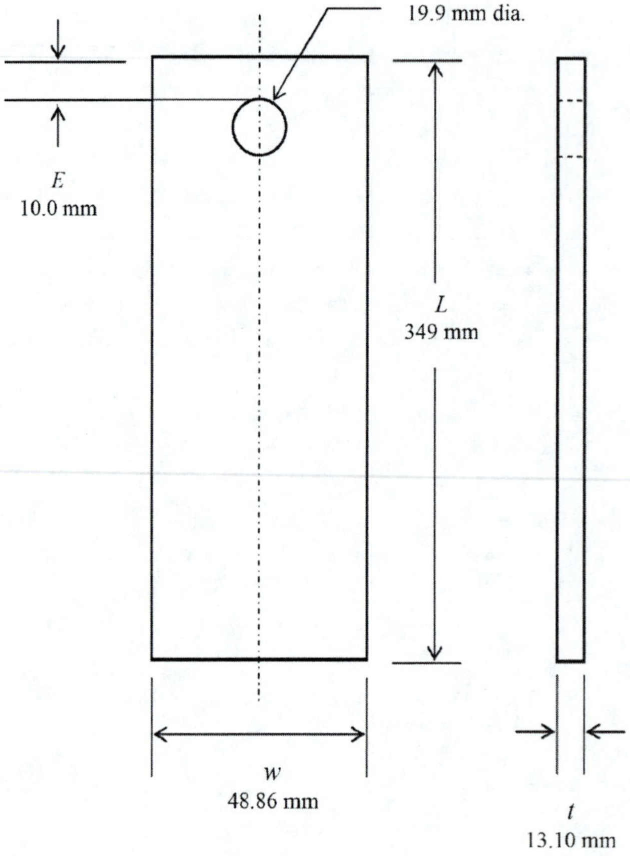

19.9 mm dia.

E
10.0 mm

L
349 mm

w
48.86 mm

t
13.10 mm

Figure 1. Schematic of the pendulum used in this experiment

Note: not to scale.

5

CHAPTER 1.7

EXAMPLE OF A LONG NARRATIVE REPORT IN LETTER FORMAT

Letter-format reports are usually used when the addressee is a client of your firm. Letter reports should be short, and they should describe projects that are small and tightly focused.

Submission date and Report title are specified.

Letter reports define author(s) and supervisor(s) in separate blocks. The supervisor is the addressee. Authors may be defined in the closing.

Background information is used to define the project task and any related work

Run-in headings save space in short reports

In short reports, displays are often attached as appendices

January 15, 2014

Professor S.M. Jeter
Instructor for ME 4053 Engineering Systems Laboratory
The Woodruff School of Mechanical Engineering
555 Engineering Drive
Atlanta, Georgia, 12345

The Motivation, or Need for the work, is defined

Subject: **Investigation of an Oscillating Water Jet in Air**

Dear Dr. Jeter,

Introduction. The task assigned to this team was to characterize the behavior of an oscillating water jet in air. Because oscillating planar liquid jets may be used for critical applications such as first wall protection in proposed inertial confinement fusion reactors, it is important to use experimental data to validate existing models of water jet behavior. The goal of this project was to assess a water jet's stability under a specific operating condition. To this end, the large scale kinematics of an oscillating planar water jet were observed in the Thermal Hydraulics Laboratory of the Georgia Tech School of Mechanical Engineering on January 10, 2014. Specifically, the maximum transverse deflection of the jet was measured at various positions below the nozzle. The measured values of these deflections were found to agree well with predictions based on a simple ballistic trajectory model.

Apparatus. The observations were performed using the oscillating jet facility in the Mechanical Engineering Laboratory, Room 3317 of the Woodruff School's MRDC building. A schematic diagram of this oscillating jet is shown in Figure 1. Figure 2 illustrates the oscillating water jet in operation before a visual measurement grid calibrated in centimeters. The jet is generated by water flowing from a rectangular nozzle that is 1 cm wide by 10 cm long. The nozzle is oscillated by a Scotch yoke mechanism, driven by a variable speed DC motor. The magnitude of the oscillation is adjusted by changing the eccentric on the motor shaft, and the frequency is adjusted by altering the DC voltage. In this experiment, the zero-to-peak oscillation amplitude was 5.4 mm, and the frequency was 6 Hz.

Procedure. The transverse deflection of the water jet was measured by imaging the jet at various positions below the nozzle with a consumer VHS camera operating at the standard 30

1

frames per second (FPS). The nozzle exit velocity was determined using a Dantec Type 55X Modular LDA single axis LDV. This exit velocity agreed well with the average velocity calculated from the volumetric flow.

To obtain the width, W, of the jet at the various positions downstream of the nozzle, a video image of the jet was obtained while the nozzle was kept stationary. The video image was then displayed on the monitor. The width of the jet image was measured directly on the screen, and the ratio of the screen size to the actual size was calculated. To determine the maximum transverse deflection of the jet, video imagery of the complete oscillating jet was obtained. This video was then displayed frame by frame, and the points of maximum deflection on both edges of the jet were identified. The distances between these points were measured directly on the monitor screen and were corrected using the calibration ratio in order to give an accurate value for the width, E, of the envelope of the trajectory of the jet. Using this value of E and the previously measured value of W, the maximum deflection, δ, of the jet was calculated using Equation 1

It is acceptable to define variables in advance of the symbolic equation display

$$\delta = \frac{E - W}{2} \tag{1}$$

When Equation 1 is used, the calculated deflections correspond to motion of the jet's centerline. This process of measurement and calculation was repeated for each downstream position that was monitored.

Results. With the jet oscillating at 6 Hz and a zero-to-peak amplitude of 5.4 mm, the maximum deflections were determined at five positions ranging from the nozzle exit to 0.9 m downstream. The resulting data are shown in Table 1 along with corresponding predictions from a model based on a simple ballistic trajectory for the jet. The raw measurement data and the calculations for Experimental δ, for Model δ and for error are collected in Attachment 1.

2

Table 1. Experimental data, error and model

y	E	W	Exp δ	error	Model δ
m	mm	mm	mm	mm	mm
0.0	21	13	4	0.7	5
0.2	34	14	10	0.8	12
0.4	51	15	18	0.9	20
0.6	65	16	24	0.9	28
0.9	90	17	36	1.6	39

In data presentations, discrepancies and irregularities should be explained

As shown in Table 1 and in the graphical presentation of Figure 3, the experimental data agree well with a simple ballistic model. At each measurement point, the experimental data values are below the model's prediction; however, the upper limits of the error calculations are close to the model's predicted values. The observed discrepancy between measured and predicted values could represent some higher order dynamic effect that was not considered in the calculation. However, this discrepancy could also represent a data collection problem; image data was collected at only 30 FPS, and this may have been too slow to detect the very transient maximum deviation of the jet. Overall, the results show that the oscillation of a planar liquid jet in air is stable and that such jets can be useful in critical applications.

A polite statement concludes the technical presentation and returns to the letter format. This allows the authors to use a first-person style for the closing sentence.

With this report, our investigation is concluded. We thank you for the opportunity to examine this system, and we will happily answer any questions that you may have.

Sincerely,

H.C. Elwell
Undergraduate student
The George W. Woodruff School of Mechanical Engineering

G.P. Burdell
Undergraduate student
The George W. Woodruff School of Mechanical Engineering

3

Author affiliations are displayed in the signature block of this report. When corporate letterhead is used, these affiliations are displayed on the first page of the letter.

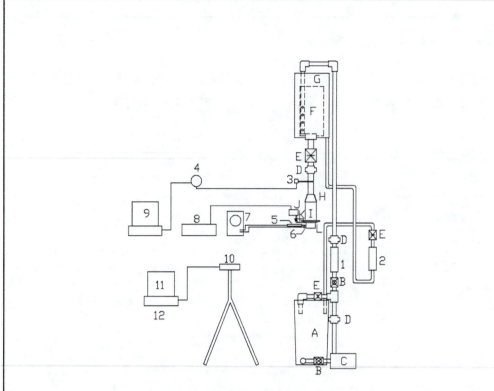

Figure 1. Schematic of the Georgia Tech oscillating jet system

	Functional Components		Instrumentation
A	Reservoir Tank	1	Flowmeter, 0-80 GPM
B	Ball Type Flow Control Valve	2	Flowmeter, 0-5 GPM
C	Centrifugal Pump	3	Pitot Tube
D	Pipe Union	4	Pressure Transducer
E	Butterfly Type Flow Control Valve	5	Linear Variable Displacement Transducer (LVDT)
F	Constant Head Tank	6	Accelerometer
G	Overflow Tank	7	Storage Oscilloscope
H	Rubber Bellows	8	Variable DC Power Supply
I	Flow Conditioner Assembly	9	Data Acquisition System
J	DC Oscillator Motor	10	Camera
K	Scotch Yoke Mechanism	11	Video Monitor
L	Nozzle	12	VHS Player/Recorder

4

Word labels are usually best for drawings and diagrams that will be delivered to clients. However, some complex displays are difficult to label clearly with text. In these circumstances, alphabetical or numerical labels are accepted so long as a legend is placed adjacent to the diagram, as here.

Photographs, like diagrams and drawings, should provide labels for important features

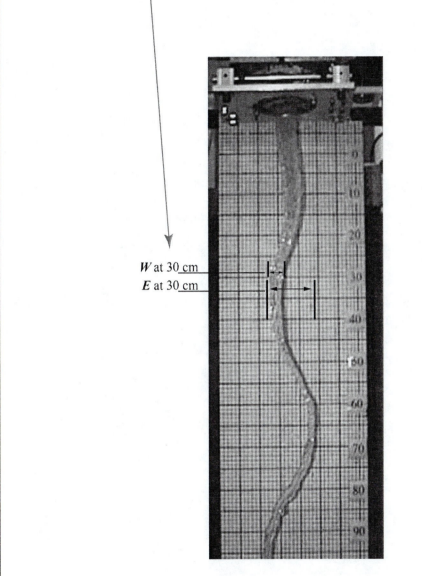

W at 30 cm
E at 30 cm

Figure 2. Oscillating water jet with measurement values illustrated

5

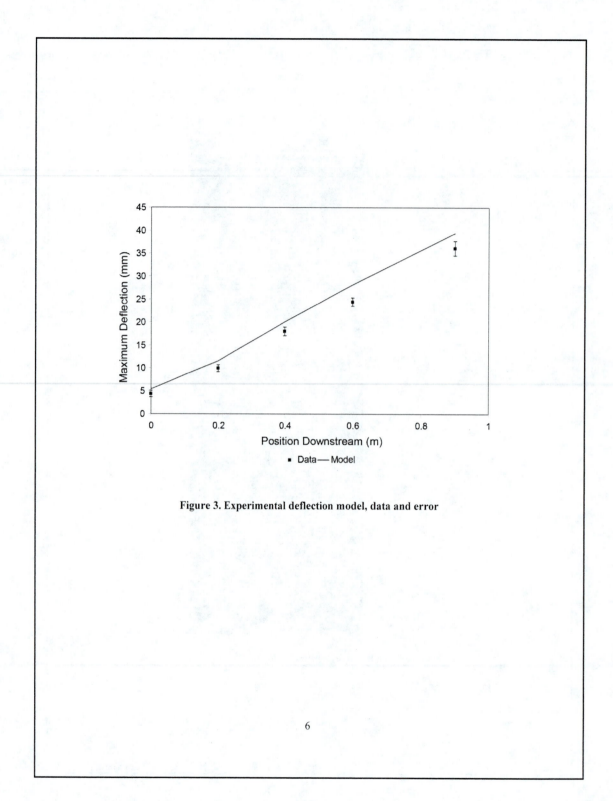

Figure 3. Experimental deflection model, data and error

6

Attachment 1. Experimental Data and Predictions for Liquid Jet

file: OscJet00   SMJ, 6 Jan 2014
Experimental data and simple model for envelope of oscillating jet.

Parameters and unique data

   59 = measured flow rate (GPM)   0.0037 = meas flow rate (m^3/sec)
  3.80 = LDV measured nozzle exit vel (m/sec)   0.0010 = nozzle cross-section area (m^2)
  3.72 = calculated nozzle exit vel from FM(m/sec)   9.81 = grav accel (m^2/sec)
  3.80 = nozzle exit velocity used (m/sec)
  6.00 = frequency (Hz)   0.00537 = amplitude (m)
 37.70 = angular velocity (rad/sec)

Deflection calculated from ballistic trajectory

y	t	th-max	x-0	vel-0	model delta
m	sec	rad	m	m/sec	m
0.00	0.0000	1.5708	0.0054	0.0000	0.0054
0.20	0.0495	0.4922	0.0025	0.1784	0.0114
0.40	0.0939	0.2754	0.0015	0.1948	0.0198
0.60	0.1345	0.1947	0.0010	0.1986	0.0278
0.90	0.1902	0.1386	0.0007	0.2005	0.0389

Deflection calculated from experimental data

y	E	E	W	W	exp delta	STD E	STD E	STD W	STD W	2 x STD delta
m	in.	m	in.	m	m	in.	m	in.	m	m
0.00	0.837	0.021	0.493	0.013	0.0044	0.012	0.00030	0.025	0.00064	0.00070
0.20	1.328	0.034	0.545	0.014	0.0099	0.020	0.00051	0.022	0.00056	0.00076
0.40	1.993	0.051	0.577	0.015	0.0180	0.022	0.00056	0.030	0.00076	0.00094
0.60	2.548	0.065	0.627	0.016	0.0244	0.030	0.00076	0.019	0.00048	0.00090
0.90	3.530	0.090	0.688	0.017	0.0361	0.054	0.00137	0.031	0.00079	0.00158

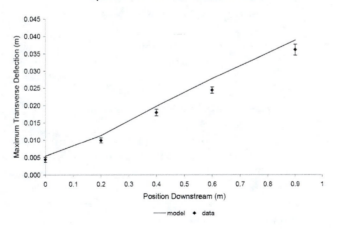

Comparison of Deflection Model with Data

7

CHAPTER 1.8

EXAMPLE OF A LONG NARRATIVE
REPORT IN MEMORANDUM FORMAT

This is an open-ended project. In reports on such projects, the Objective and Motivation must be explicitly stated

The object under study is described fully

The experimental problem is formulated

The method of analysis and the expected results are outlined

TO: Instructor   23 November 2014

FROM: C. C. Pascual and S. M. Jeter

SUBJECT: Measurement of Heat Leak from the Copper Cylinder Used in Convection Heat Transfer Experiment

Introduction

The objective of this project was to evaluate the conduction heat leakage from the heated cylinder that is currently used in the convection heat transfer experiment located in the Thermal Systems Instructional Lab of the George W. Woodruff School of Mechanical Engineering. The copper cylinder is heated by an embedded electric resistance heater and cooled by a cross flow of air. The power delivered to the resistance heater can be accurately measured; however, some heat escapes from the cylinder by radiation or by conduction, either through the supports or through the power and instrumentation leads. This extraneous heat transfer is called a heat leak, and this heat leak should be subtracted from the electric power to yield the convection heat transfer rate.

Heat leak measurements were conducted on 9 and 10 November 2013. The heat leak from the cylinder was measured at three different air speeds by low power steady-state calorimetry. The heat leak was found to be large and to depend on the air speed. A regression model for the heat leak conductance was developed for use in correcting the measured electrical power for conduction losses. The experimental heat leak was found to be higher than would be expected if one-dimensional conduction through the supports were the only source of heat loss. The higher value is attributed to conduction through the electrical leads in addition to conduction through the supports. In addition, the supports and one lead are exposed to the air stream, causing them to act as fins; this is the source of the observed dependence on air speed in the experimental results. Using the measured results, a regression model for a heat leak coefficient was developed. This model can be used in the future to compute an improved estimate of the true convection heat rate.

Apparatus and Procedure

Before collecting data for this task, it was necessary first to insulate the existing cylinder against convection. Data was then collected by applying an electric current to the heater and

1

For open-ended projects, test conditions and assumptions must be fully defined

recording the steady state temperature of the cylinder. At steady state, it is assumed that all the input power is escaping through the conduction heat leak paths.

To insulate the heated cylinder, it was first installed in an Aerolab subsonic induced draft wind tunnel. This wind tunnel has a constant speed fan and a bypass gap to adjust the air speed through the test section. To eliminate convection and to block radiative heat leak, the heat transfer surface of the cylinder was covered with two layers of 25 mm (nominally 1 inch) thick pre-formed glass fiber pipe insulation. With the convection surface well insulated, almost all steady state power can then be assumed to be conducted away by the structural supports of the cylinder and by the power and instrumentation leads.

After the cylinder was insulated, the power leads were connected to a Weston model 310 wattmeter and then to an autotransformer that was used to adjust the AC voltage applied to the resistance heater. Measurements of the cylinder's average surface temperature were obtained by using fifteen type T thermocouples embedded in the insulation near its surface. These thermocouples were connected to a Metrabyte EXT-16 submultiplexer board with internal data acquisition that includes a high resolution, 12-bit analog to digital converter. One additional type T thermocouple, also connected to the data acquisition system, was used to monitor the free stream air temperature. Finally, a TSI Velocicalc model 8350 portable thermal anemometer was used to measure the free stream air speed.

To take heat leak measurements, the wind tunnel fan was activated, and the air speed was adjusted, using the bypass gap, to take measurements at three representative air speeds, 14, 9, and 4.5 m/sec. The resistance heater was energized at a relatively low power, either 5 or 10 Watts, and the cylinder surface temperature was monitored until a steady state was achieved. The experiment was hindered by the long time, 2 to 3 hours, required for the insulated cylinder to reach a steady state. When a steady state was achieved, the average surface temperature, the free stream temperature, and the free stream air speed were measured and recorded.

2

Results

The complete set of data and calculations is provided on the accompanying spreadsheet, Attachment 1. The essential data and results are summarized in Table 1 and illustrated in Figure 1.

Table 1. Heat leak data and calculated conductance

Air Speed	Electrical Power	Air Temp	Surface Temp	Experimental UA	Regression UA	Error
m/sec	W	C	C	W/C	W/C	%
13.9	5.0	19.0	34.3	0.327	0.331	1.4
9.0	5.0	16.6	36.2	0.255	0.258	1.1
9.0	10.0	17.4	54.5	0.270	0.257	-4.6
4.5	5.0	18.1	45.0	0.186	0.191	2.7

Table 1 includes the heat leak conductance, *UA*, which was computed using Equation (1)

$$UA_L = \frac{\dot{W}_E}{T_S - T_\infty} \qquad (1)$$

where, UA_L is the heat leak conductance (W/C), $\dot{W}_E$ is the electrical power (W), T_S is the surface temperature (C), and T_∞ is the free stream air temperature (C). The tabulated experimental conductances show two notable features. First, the conductance increases with air speed. This behavior is not compatible with simple one-dimensional conductance. A 73% increase in the conductance is associated with the overall 440% increase in air speed. This increase is more typical of convection from a fin rather than mere one-dimensional conduction to an ambient heat sink. Second, the heat leak is large, on the order of 20% of the convection heat rate from the cylinder. The heat leak's absolute size is attributed to excessively conductive supports and, especially, the leads. The relatively small convection heat rate, caused by the short length and correspondingly small convection surface of the test cylinder, enhances the relative significance of the conduction heat leak.

The experimental bias was estimated by error propagation analysis following the method of Kline and McClintock [1]. The details are given in the spreadsheet included as Attachment 1. The result is an estimated limit of bias of ± .021 W/C. This value is less than 9% of the mid range value of the heat leak conductance.

3

An IEEE-Style reference citation

Note that the author's name(s) can be used as part of the citation

Figure 1 displays the experimental data with a linear regression model for the heat leak conductance data as a function of wind speed. The model is expressed in dimensional terms by Equation 2 for the wind speed, V, in m/s,

$$UA_L = 0.1244 \, \text{W/C} + \left(.01483 \frac{\text{W/C}}{\text{m/s}} \right) V \qquad (2)$$

Figure 1 demonstrates that the model represents the data well. The largest deviation of the model from the data is only -4.6%. Overall, the R-squared value is almost 98%, indicating excellent agreement between the data and the model. The model has an alpha risk of only 1%, implying a negligibly small probability that the observed correlation between the conductance and the air speed could be due to chance. This alpha risk is well below the conventional upper limit of 5%and indicates that the regression model is statistically significant.

To give perspective on the model represented by Equation 2, a quadratic model was also developed and evaluated. As detailed in the attachment, this model returned a slightly higher R-squared value, but it had an entirely unacceptable alpha risk of 54%. This high alpha value indicates that the enhanced fit is an adjustment to random error in the model; consequently the quadratic term is statistically insignificant and should not be used.

The linear regression analysis also returned a value of only 0.010 W/C for its standard error of estimate. To illustrate the tightness of the regression estimate as implied by this standard error, an experimental error band with the half-width of two standard errors of estimate is also plotted on the figure. Obviously, all of the experimental data are either within or very near this error band, and this is another indication that the model represents the data well and that the data includes no suspicious outliers.

4

Blank spaces such as this can be filled by moving text from below the figure on the next page, by resizing the figure to fit the available space, or by placing the figure above the citation to equation 2.

When a figure is separated from its text presentation, as here, readers often find it difficult to understand the explanation. Try to format your reports so that figures and explanations are on the same page.

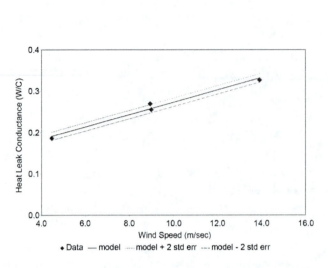

Figure 1: Heat leak conductance data, model, and error band

Results are most useful when they are compared to predictions or to established benchmarks

Discussion

Comparison with Simple Models

The experimental heat leak conductance results can be most easily described as a case of one-dimensional conductance through the supports. The supports are made of acrylic polymer. One is a cylindrical solid, and the other is a hollow cylinder to accommodate the electrical leads. Table 2 gives their dimensions and their computed conductances. Together the two supports have a one-dimensional conductance of no more than .0073 W/K, which is less than 4% of the smallest observed heat leak.

Table 2. Simple conductance calculations for the solid and the hollow supports

	Units	Solid	Hollow	Sum
Inner Dia	mm	0	32	
Outer Dia	mm	32	44	
Area	mm^2	792	760	
Length	mm	41	0	
Conductance	W/C	0.0038	0.0034	0.0073

The much larger observed value of heat leak conductance is attributed to the instrumentation for this experiment. The cylinder is equipped with 16 pairs of 24 gage thermocouple leads, one pair of 12 gage power leads, and a 12 gage grounding conductor. A

5

single 12 gage copper lead effectively 50 mm long has a conductance of .026 W/C, and a 24 gage lead has a conductance of .0016 W/C. In total, the electrical leads could allow a heat leak conductance of at least .104 W/C, which is 54% of the lowest observed value. In addition, the supports and the grounding lead are directly exposed to the air stream and are sure to experience some lateral conduction and convection heat loss. Furthermore, the power leads are connected directly to the resistance heater, which is much warmer than the convection surface itself, exacerbating the heat leak by this path. These effects are believed to account for the enhanced conduction heat leak; the fin effect in particular explains the unexpected dependence on the wind speed.

Here the author presents the conclusion that he draws from the data

Radiation Heat Leak

In addition to the observed convection heat leak, in normal operation the cylinder is cooled by radiation. However, the cylinder is not provided with a low emittance coating, so its radiation heat loss is not negligible. For nominal temperatures of 50° C at the surface and 25° C ambient, and assuming 0.3 surface emissivity, the cylinder's radiation heat leak is calculated to be around 0.04 W/C. This calculated radiation heat leak is somewhat more than 15% of the nominal conduction heat leak. This additional leak is large enough to be significant and should also be subtracted from the input power when computing the convection heat rate.

Conclusions

The conductance heat leak from the forced convection test cylinder was measured and a regression model was developed. For the expected range of wind speeds, the heat leak conductance is significant, ranging from 0.19 to 0.33 W/C. In addition, the radiation heat leak can be estimated to have a nominal value of about 0.04 W/C, also an appreciable contribution to the heat transfer from the cylinder. Proper correction of the input power for the conduction and radiation heat leaks should significantly improve the accuracy of the forced convection experiment.

IEEE format is used for citation and for reference entry

Reference

[1]   S. J. Kline and F. McClintock, "Describing uncertainties in single-sample experiments," *Mechanical engineering*, vol. 75, pp. 3-8, 1953.

6

Attachment 1. Spreadsheet for Heat Leak Experiment

file: FCCAL3   20 Nov 14 (SMJ)
Experimental determination of heat leak from forced convection source
Data collected 8 November 2014.

Summary:

0.010 = alpha risk	0.021 = error limit (W/C)
0.010 = R-squared	8.123 = percent error limit

Constants and parameters:
0.3048 = m/ft
1609.3 = m/mile

Data and calculations:

speed FPM	speed m/sec	power W	t-amb C	t-sur C	UA W/C	model UA W/C	error %	model + std err	model - std err
2745	13.9	5.0	19.0	34.3	0.327	0.331	1.4	0.341	0.321
1773	9.0	5.0	16.6	36.2	0.255	0.258	1.1	0.268	0.248
882	4.5	5.0	18.1	45.0	0.186	0.191	2.7	0.201	0.181
1763	9.0	10.0	17.4	54.5	0.270	0.257	-4.6	0.267	0.247

Linear Regression output:		Quadratic Regression output:		
Constant	0.124391	Constant	0.092897	
Std Err of Y Est	0.010105	Std Err of Y Est	0.010752	
R Squared	0.979721	R Squared	0.988519	
No. of Observations	4	No. of Observations	4	
Degrees of Freedom	2	Degrees of Freedom	1	
X Coefficient(s)	0.014833	X Coefficient(s)	0.022650	-0.000422
Std Err of Coef.	0.001509	Std Err of Coef.	0.009073	0.000482
t-stat	9.829773	t-stat	-0.875413	
alpha	0.010191	alpha	0.542230	

Uncertainty analysis:

nom Ts =	36.2 C				
nom Ta =	16.6 C				
nom W =	5.0 W				
nom UA =	0.255 W/C				
	u (units)	dUA/dx	(u dUA/ /dx)	(u dUA/ /dx)^2	
T (air)	0.733 C	0.0130	0.0095	9.11E-05	
T (surface)	0.189 C	0.0130	0.0025	6.07E-06	
Power	0.063 W	0.0510	0.0032	1.02E-05	
		u in UA	0.0104 W/C		
		error limit	0.0207 W/C		
		error limit	8.1 percent of nominal UA		

7

CHAPTER 1.9

EXAMPLE OF A CAPSTONE REPORT

The Generic School of Mechanical Engineering
Standard Institute of Technology, Collegeville GA 30332-0405
Undergraduate Instructional Laboratories

TO: Undergraduate Laboratory Instructor   11 April 2015
FROM: Undergraduate Laboratory Students in Group A03
SUBJECT: Transmittal of Final Report

Dear Professor Burdell:

This cover note defines: 1) What is delivered, 2) the main points, 3) the action item or takeaway point

We have enclosed with this letter the final report of our project to develop an improved water flow system for the LDV experiment in the undergraduate lab. Our report presents an analysis of the old, unsatisfactory system, our design for the new system, and the results of our validation tests on the redesigned system. The redesigned system was found to be functional, but numerous problems remain; our report briefly presents the existing problems, with our analysis and recommended course of action.

We thank you for your encouragement and assistance throughout this project.

Sincerely,

Ernst T. Clark, Group Leader

All team members sign the cover letter

Wileng A. Scholar

Claire L. Scribner

Attachment: Final Report

53

(This page intentionally left blank.)

This formal cover sheet identifies all the people involved in the project, their roles and their affiliations

The authors take responsiblity for all the work presented in the report

The advisor, supervisor or client employs the authors, assigns their tasks, and approves their work

The submission date enables you to keep track of versions, and it prevents disputes about on-time delivery of results

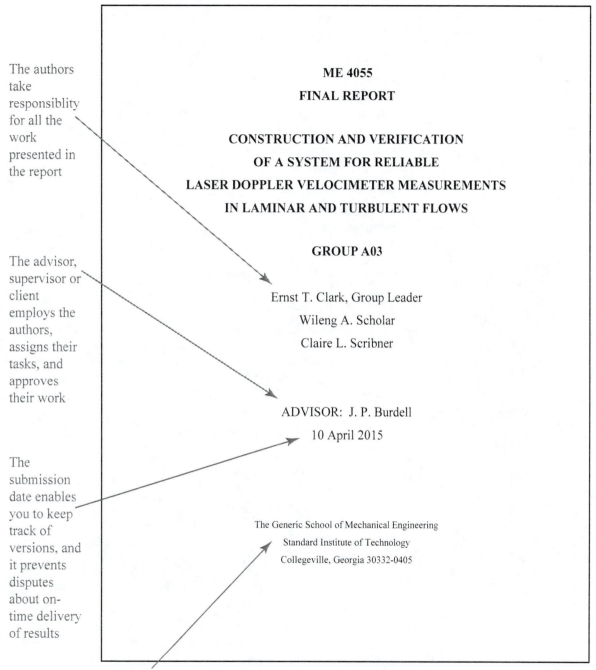

ME 4055

FINAL REPORT

CONSTRUCTION AND VERIFICATION

OF A SYSTEM FOR RELIABLE

LASER DOPPLER VELOCIMETER MEASUREMENTS

IN LAMINAR AND TURBULENT FLOWS

GROUP A03

Ernst T. Clark, Group Leader

Wileng A. Scholar

Claire L. Scribner

ADVISOR: J. P. Burdell

10 April 2015

The Generic School of Mechanical Engineering

Standard Institute of Technology

Collegeville, Georgia 30332-0405

The authors' affiliations may also be presented in letterhead

A **Table of Contents** is useful for any report that runs to ten pages or more

TABLE OF CONTENTS

i

A well-
structured
Abstract
defines:

1) Need

2) Specific
objectives

3) Methods

4) Results
and
significance
of results

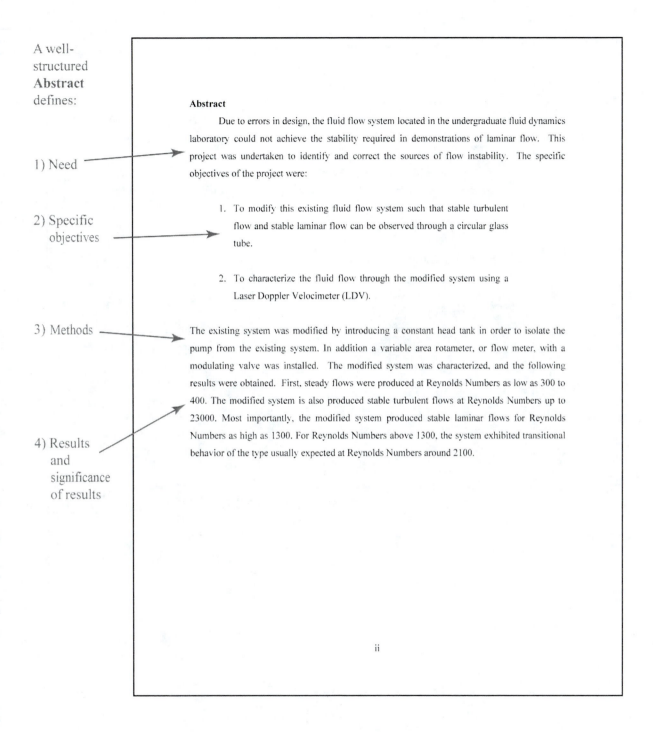

Abstract

Due to errors in design, the fluid flow system located in the undergraduate fluid dynamics laboratory could not achieve the stability required in demonstrations of laminar flow. This project was undertaken to identify and correct the sources of flow instability. The specific objectives of the project were:

1. To modify this existing fluid flow system such that stable turbulent flow and stable laminar flow can be observed through a circular glass tube.

2. To characterize the fluid flow through the modified system using a Laser Doppler Velocimeter (LDV).

The existing system was modified by introducing a constant head tank in order to isolate the pump from the existing system. In addition a variable area rotameter, or flow meter, with a modulating valve was installed. The modified system was characterized, and the following results were obtained. First, steady flows were produced at Reynolds Numbers as low as 300 to 400. The modified system is also produced stable turbulent flows at Reynolds Numbers up to 23000. Most importantly, the modified system produced stable laminar flows for Reynolds Numbers as high as 1300. For Reynolds Numbers above 1300, the system exhibited transitional behavior of the type usually expected at Reynolds Numbers around 2100.

ii

Introduction:
Need and
Problem
stated in
general terms

Background:
Specific
description
of existing
system and
conditions

The goal
is clearly
specified

Introduction

The objective of this project was to design, construct, and test a very stable water flow system for use in undergraduate experiments that use an LDV system to measure fluid flow. The task of this flow system was to produce stable laminar and turbulent flows in a circular glass measurement pipe over a wide range of volumetric flow. This system was to be developed by modifying an existing fluid flow system that could not be regulated well enough to develop stable laminar flow. To complete this project successfully, it was necessary to identify the problem with the existing system, to design and fabricate modifications for this system and to experimentally verify that those modifications meet the specified requirements of stable laminar and turbulent flow over a specific range of Reynolds numbers. The steps of problem identification, redesign and experimentation are presented in the next sections.

Background

A schematic of the previously existing LDV and flow system is shown in Figure 1. This flow system uses a single pipe to deliver water from a pump to a clear pipe through which the flow rate is obtained at a measurement point. From this point the flowing water returns to the reservoir that houses the pump. This system was developed for use in class demonstrations and for data collection in undergraduate laboratories. In order to be a useful teaching tool, a system such as this must reliably produce stable flow. However, flow in this device was found to be so unstable as to render the device unsatisfactory; it was removed from student use pending modifications. Before redesign commenced, the desired performance of the modified system was defined as follows: it must produce stable laminar flow at Reynolds Numbers between 400 and 1300, stable turbulent flow at Reynolds numbers between 2200 and 23000, and transitional flow behavior at Reynolds numbers between 1400 and 2100.

1

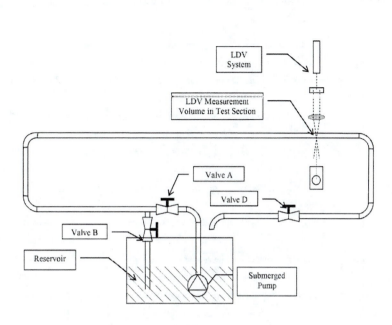

Figure 1. Original fluid flow system, showing locations and integration of LDV system, test section, reservoir, pump, and flow control valves

Problem Identification. The flow variations of this system were found to be caused by dynamic interaction between the pump and the piping system. This was attributed specifically to the close integration of the pump, which was connected directly to the pipe loop containing the glass measurement pipe. This direct connection caused erratic velocity profiles, particularly at the low Reynolds Numbers typical of laminar flow. The pump further disrupted the fluid flow by introducing mechanical vibrations into the measurement pipe and the return tube. The existing flow control valve also presented some problems; it may have contributed to turbulence in the fluid flow, and it was difficult to precisely adjust low flow rates using the existing valve.

Redesign. To address these problems, the system was redesigned to isolate the pump from the flow pipes. The modified system is shown in Figure 2. A constant head tank and a return tank were added to the system between the pump and the measurement pipe. In the modified system,

2

Full problem
statement

Technical
terminology
is appropriate
here, even
though little
is required for
this project

This excellent figure description is both brief and complete

the measurement tube receives fluid from the head tank, while water that overflows the head tank is returned to the reservoir through a newly added return tube. The water flow to the head tank is regulated through manual adjustments to valves A and B. This modification eliminated pump surge by allowing for a consistent fluid head, which is provided by the difference of fluid levels in the head tank and the reservoir. In design of this system, the height of the head tank required careful consideration. This tank had to overcome head loss due to frictional forces from the pipelines, valves, and fittings throughout the system. However, the height of the head tank was constrained by the limited pump head. Additionally, the return tube was a possible source of fluctuation, and the system had no flow meter to allow direct reading of the flow rate.

Fluid velocity in this system is measured in a glass measurement pipe downstream from the new constant head tank. Measurements are taken with a Laser Doppler Velocimeter, or LDV, which crosses laser beams through the glass pipe at a measurement point where the flow is fully developed. The LDV, shown in Figure 3, operates on the differential Doppler principle. A monochromatic helium-neon laser beam is sent through a beam splitter to produce two coherent beams. These beams are then directed by a lens to cross at the test section inside the circular glass measurement pipe. When these beams cross, they produce a pattern of light and dark fringes that are separated by a known distance D_f. As entrained particles cross through each bright fringe, a pulse of scattered light is focused into a mirror, which reflects the light pulse into a photomultiplier tube (PMT). A data processing unit connected to the PMT detects the fringe passing frequency from these pulses. The velocity of the fluid is calculated from the known distance, D_f, between the fringes and the measured frequency.

A rotameter, or variable area flowmeter, was added to the system to allow direct measurement of the liquid flow rate. The measurements obtained from this meter can verify the calculated flow rate, thus validating the collected LDV data. The rotameter itself can be calibrated by the weighing tank method as described below. Future experiments can compare the volumetric flow calculated from LDV data with direct measurement of the flow obtained using the rotameter.

3

Full-page landscape displays, which are wide but not tall, may be rotated for display in a report. When you do this, the bottom of the display should be oriented to the reader's right side.

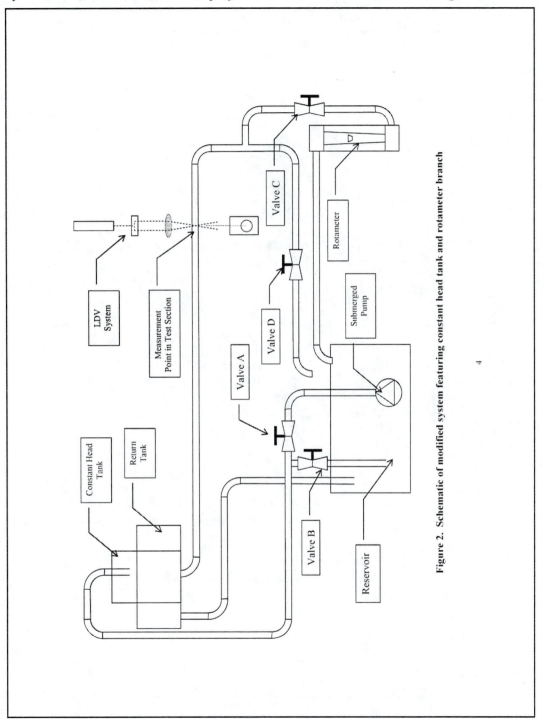

Figure 2. Schematic of modified system featuring constant head tank and rotameter branch

4

A large, portrait-oriented image can be displayed in normal orientation. When you have several large images to display in sequence, consider placing them in an appendix, to avoid disrupting the running pages of the report.

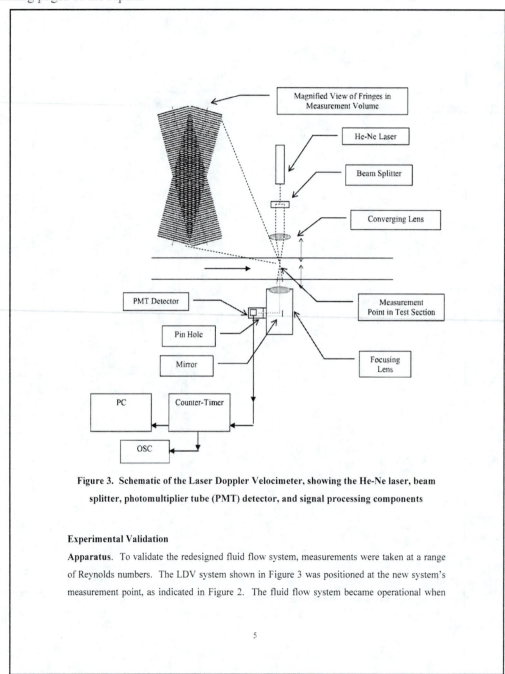

Figure 3. Schematic of the Laser Doppler Velocimeter, showing the He-Ne laser, beam splitter, photomultiplier tube (PMT) detector, and signal processing components

Experimental Validation

Apparatus. To validate the redesigned fluid flow system, measurements were taken at a range of Reynolds numbers. The LDV system shown in Figure 3 was positioned at the new system's measurement point, as indicated in Figure 2. The fluid flow system became operational when

5

This section provides a good description of a small experiment

This is an orderly description of a small experimental procedure:

1) Task goal

2) Functional need

3) Experimental action

the pump was activated, sending water from the reservoir tank to fill the constant head tank. During operation, water flows from the head tank into the glass measurement tube, where its velocity is measured with a TSI System 9100 LDV. The water then continues through the return pipe to the reservoir tank.

Procedure. To determine the maximum velocity of the water flow, the LDV must obtain a velocity profile across the glass measurement pipe by taking a sequence of measurements at discrete points across the pipe. The first measurement in such a profile was obtained at a near wall reference point calculated to be about .43 mm (0.017 inch) from the tube wall. To locate this point, the laser traversing table was moved towards the near tube wall, taking measurements until a low particle count was obtained. A dial indicator attached to the adjusting table was then set to zero, and the adjusting table was moved a distance of 27.9 mm (1.09 inch) from the first point, which should place it in the desired position about 0.43 mm from the far wall. The flow rate was noted, and the procedure was repeated until similar rates were counted at both sides of the tube, indicating that a symmetric alignment had been found. Once this adjustment was completed, the LDV measurement volume was centered so the flow rate could be adjusted.

The LDV output to the computer, as shown in Figure 3, was used to measure the maximum velocity of the water at the center of the tube. The flow rate was adjusted to the desired velocity using the flowmeter throttling valve. The pump throttling and the bypass valves were adjusted to a flow that just overflowed the overflow tank without reducing the flow through the pump to an unacceptably low rate.

The laser was reset to the near wall location .43 mm from the tube wall, and the average velocity, standard deviation and turbulent intensity were measured. This data was entered into a spreadsheet to be used for further analysis. Once the data for a point was recorded, the laser was moved 1.3 mm (0.050 inch) towards the other wall, and another measurement was taken. This data collection was repeated until measurements had been taken across the whole of the tube. At that time the volumetric flow and the actual Reynolds Number were calculated using a spreadsheet, and the velocity profile was plotted.

6

The rotameter, or flow meter on the right side of Figure 2 was introduced to the system to independently determine the volumetric flow rate of the fluid in the measurement pipeline. This flow meter must be calibrated so that its scale will correlate to known flow rates. This calibration was performed using the weighing tank method. A digital stopwatch and a Toledo Type 211795 truck scale were used.

For calibration, the rotameter was set up to flow water into the weighing tank. Its scale is read as numbers from 0 to 100 in increments of 10. When fluid runs through the meter, a float inside indicates a number on this scale to indicate the fluid flow rate. The weight of the water inside the tub changes as fluid leaves the rotameter and enters the tub. This changing water weight was recorded at the beginning and end of a time interval lasting between 3 and 5 minutes. This process was repeated for each desired increment on the scale of the rotameter.

The resulting change of mass in kilograms of water was converted to a change of volume in liters; it was then divided by the corresponding change in time in seconds to give a volumetric flow rate. The measured values of volumetric flow rate (L/sec) were plotted against the scale numbers of the rotameter. Regression analysis gives this calibration equation,

$$Q_{corr} = 0.0418\,X + 0.044\,\frac{\text{liter}}{\text{min}} \qquad (1)$$

where Q_{corr} is the volumetric flow rate in L/sec and X is the scale number. The R^2 value for this calibration was 0.9994, and the Combined Uncertainty of the calibration was 0.21 L/min. These results are completely acceptable for this application. This equation can be used to determine the corrected volumetric flow rate that corresponds to the scale number indicated on the flowmeter.

Results

The objective of this project was to design a fluid flow system capable of producing stable laminar flow as well as stable turbulent flow, and the measurement results indicate that this objective was accomplished. With the addition of the head tank, the modified system produces a laminar parabolic flow profile for the first time, as shown in Figure 4. The flow

7

velocity profile in the figure for $Re = 380$ closely resembles the laminar model, which is based on Equation (2),

$$u/U_c = 1 - \left[\frac{r}{R}\right]^2$$
$$U_c = 2u_{ave}$$

(2)

Some error is still apparent. Poor adjustment in the measurement locations may have caused some shift in the measurements. However, the data in Figure 4 still give adequate but not completely identical agreement with the model. Even with the residual error, the modified system is a great improvement over the previously existing system, which never produced even an approximately parabolic flow profile.

Good plots can compare numerous data sets and models

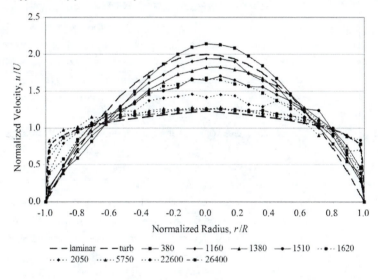

Figure 4. Flow profiles obtained with the modified system over a range of Reynolds Numbers

Detailed explanation of analytical results

The range of this laminar flow for the modified system was also determined. As the speed of the fluid approaches the transition zone (*i.e.*, Reynolds Number 2100 to 4000), the flow profiles grow less stable, yielding curves that are not smooth, as shown by the curve for $Re =$

8

Anomalous
data is noted
and discussed

1510 in Figure 4, for example. For this system, the velocity profile begins to show some fluctuation or instability as low as $Re = 1500$. This early apparent transition to turbulence may be due to some disturbance induced in the flow by the piping system and / or some residual turbulence in the fluid when it reaches the test section. Consequently, the effective range for this system is below $Re = 1500$. An obvious challenge is to upgrade the system to generate a smooth laminar profile for Reynolds Numbers closer to 2100, which is the textbook measure of the end of the laminar zone; however, this enhancement would certainly be non-trivial.

The transition from laminar to turbulent flow was found to be complicated. The most characteristic features of this transition are the normalized centerline velocity and the corresponding standard deviation of the velocity. The normalized centerline velocity typically decreases during transition as the profile evolves into the more uniform turbulent profile. As displayed numerically in Table 1 and graphically in Figure 4, the normalized velocity in the core flow does tend to decrease as the Reynolds Number increases. Such behavior is expected; however, the decrease in this case is substantial even within the accepted laminar range as presented in all standard texts such as [1]. During these tests, the display of the active histogram of the local velocities fluctuated even when the flow rate was relatively low. The results were the uneven flow profiles displayed in Figure 4. Fluctuation is expected in the transition range, but for Reynolds Numbers below the expected 2100, and even for Re as low as 1200, minor fluctuation was observed in real time and were demonstrated in the measured profiles.

A reference
citation can
be used as a
grammatical
component
of a sentence

The monotonic decline in the normalized velocity as the Reynolds Number increases is a characteristic feature and deserves some further investigation. According to theory, the normalized centerline velocity should remain at 2.0 in the laminar range, which corresponds to a Reynolds Number somewhat below about 2100. In this regime the velocity profile should be parabolic. After transition to turbulence around a Reynolds Number of 4000, it should remain nearly constant at around 1.2 corresponding to the expected power law profile. In the transition region (*i.e.*, $2100 < Re < 4000$), there should be a decline in peak velocities. Figure 5 shows measurements of the peak velocities plotted against the logarithm of the ascending Reynolds Numbers. In the transition region, the normalized peak velocity decreased as expected.

9

Table 1. Normalized velocity and normalized standard deviation at the centerline of the test section for the experimental range of Reynolds Numbers (*Re*)

Re	U_c/U_{avg}	norm SD at CL[a]	*Re*	U_c/U_{avg}	norm SD at CL[a]
384	2.14	0.0342	2053	1.41	0.0291
1160	1.94	0.0458	2711	1.37	0.1297
1380	1.82	0.0398	5749	1.25	0.1591
1510	1.65	0.0501	22576	1.26	0.1035
1620	1.68	0.0322	26427	1.25	0.0949
1820	1.71	0.0493			

[a] The normalized Standard Deviation (SD)
of the velocity sample at the pipe centerline (CL).

The trend in the standard deviation is not as widely documented as the trend in the normalized velocity, and its behavior was found to be more complex. The normalized standard deviations of the centerline velocities are shown by the error bars on each point in Figure 5. This statistic is direct evidence of scatter in the velocity data caused by turbulent fluctuations. The scatter tends to increase as the Reynolds Number increases in the nominally laminar regime. However, rather than continuing to increase, the normalized standard deviation reaches a maximum in the transition region where the normalized standard deviation is larger than in either the laminar or the highly turbulent regions. This behavior, while not necessarily expected, is reasonable. In the transition region, the onset of turbulent eddies causes relatively large velocity fluctuations. This fluctuation increases into the mildly turbulent range. However, as the flow rate is increased even further, the velocity in the turbulent eddies clearly does not increase as quickly as does the local mean velocity. Consequently the standard deviation decreases, and the ratio called the turbulence intensity even begins to decrease. The greatest relative fluctuation is then in the transition, not the fully turbulent regime.

10

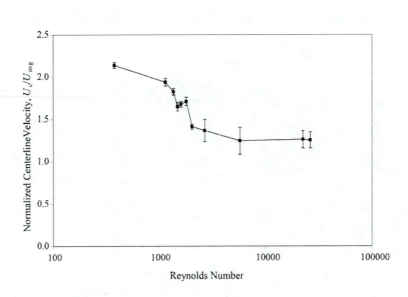

Figure 5. The normalized centerline velocities plotted against Reynolds Number accompanied by error bars 2 sample standard deviations in length to illustrate the relative turbulence

In the **Discussion** section, results are reviewed and unusual outcomes are noted

Discussion

Overall performance of the modified system is acceptable. However, at higher Reynolds Numbers, but still within the expected laminar zone, the flow profile becomes unstable and deviates from the model. Initially, this problem was thought to stem from back pressure fluctuations in the return line to the reservoir. In the original modification, the outlet of the return line was below the water level. This arrangement allowed unavoidable fluctuations in the reservoir water level and currents in the reservoir to affect flows in the test section. To eliminate this possibility, the pipe exit was raised so that water pours freely into the reservoir at ambient pressure. Some improvement in the flow stability was noted after this improvement, but fluctuations were still not eliminated.

11

The design of the constant head tank may be a source of this remaining disturbance. In order to maintain a constant head without splashing, the water must be discharged below the surface of the water in the tank. This submerged inlet may currently be so close to the exit of the tank that fluctuations in the discharge may influence the flow in the test section. In support of this hypothesis, air bubbles were observed to sometimes collect in the measuring tube. These bubbles are created by splatter from the return line at the reservoir; and they then can be induced into the pump and travel through the supply line to the head tank. At low flow rates any bubbles sent to the head tank should rise to the top and leave the system. However even at relatively low Reynolds Numbers in the upper end of the laminar flow range, some bubbles remain in the liquid long enough to be carried into the measuring tube. Indeed, the tube must be flushed of bubbles periodically. This phenomenon is exacerbated at higher Reynolds Numbers because at low laminar Reynolds Numbers the flow is slow enough to allow bubbles to escape from the free surface rather than being drawn into the measuring tube. Obviously, a larger constant head tank would be desirable, although available space now limits installation of a larger tank.

The troublesome fluctuations might be minimized by moving the discharge of the supply line into the head tank as far as possible from the drain without actually lifting it out of the water. This modification should be feasible. Alternatively, a diffuser could be attached to the hose entering the head tank. This diffuser would allow fluid to flow more smoothly from the pump into the head tank, preventing the pumping surges from directly disturbing the flow in the test section.

As future work, changes are suggested, as are alternative analyses

It is possible that some variations may derive from other sources. For example, the flow profile might not have fully developed by the time it is measured. The required entry length, l_E, which is the length of pipe necessary for full development of the velocity profile, is quoted in standard texts such as Munson [1] as

$$\frac{l_E}{D} = 0.06\,Re \tag{3}$$

where D is the diameter of the pipe, and l_E is the pipe's length. The measuring pipe is 3.0 meters long, and its diameter is 38.1 mm. Consequently, when the flow rate exceeds that necessary for

12

Reynolds Number to be 1300, the current pipe is too short for laminar flows to develop fully. The simplest way to solve this problem is to use a measurement pipe with a smaller radius. A longer pipe could also solve this problem; however, space constraints make such a pipe impractical in the current facility.

Conclusions

An existing fluid flow system was modified to allow it to reach stable turbulent flow and stable laminar flow through a circular glass tube. A Laser Doppler Velocimeter was used to characterize fluid flow through the modified system.

The modified system produced Reynolds Numbers as low as 300 to 400, it produced stable turbulent flows at Reynolds Numbers up to 23000, and it produced stable laminar flow for Reynolds Numbers as high as 1300. For Reynolds Numbers above 1300 the system exhibited transitional behavior of the type usually expected at Reynolds Numbers around 2100.

Several recommendations have been developed to address the problems of premature transitional behavior at Reynolds Numbers between 1300 and 2100: (1) the size of the head tank should be increased, (2) a diffuser should be added to the supply line, and (3) the supply line discharge into the head tank should be moved farther from the outlet. In addition, the diameter of the glass tube should be reduced to insure that the water has achieved fully developed flow at the measurement point.

Acknowledgments

We would like to thank our instructor for providing us this opportunity to work directly with Laser Doppler Velocimeter technology and to apply our knowledge of fluid mechanics and fluid systems to a real life application. We would also like to thank our teaching assistants for steering our progress in the laboratory and for providing us with much needed advice and materials. They also instructed us how to use the laboratory equipment and ensured that our work was correct.

Reference

13

Acknowledgments in reports: When you work for a client, you should describe any assistance you received from the client's staff. This is polite, and it is good for business.

If you do funded research, you should disclose all sources of funding that contributed to the work. This is usually required in order to address concerns about conflicts of interest.

[1]   B. R. Munson, D. F. Young, T. H. Okiishi, and W. W. Huebsch, *Fundamentals of fluid mechanics / Bruce R. Munson, Donald F. Young, Theodore H. Okiishi, Wade W. Huebsch*: Hoboken, N.J. : Wiley, c2009. 6th ed., 2009.

Appendix: Sample Calculations for

Determination of Reynolds Number from Average Fluid Velocity

The Reynolds Number is related to area average fluid velocity according to the following equation:

$$Re = \frac{V D}{v} \qquad (A.1)$$

where: Re is the Reynolds Number, V is the Average fluid velocity, D is the Fluid passage diameter (the pipe's inner diameter in this case), v is the Kinematic viscosity.

Representative experimentally obtained values for these variables were:

$V =$ 0.0512 m/s   (0.168 ft/s)

$D =$ 38.1 mm   (nominal 1 1/2 inch glass pipe, 1.50 in. ID)

The kinematic viscosity of water at 25 °C, representative of experimental conditions, is[1]

$v =$ 9.03×10^{-7} m²/s   (9.72×10^{-6} ft²/s)

Substituting these values into Equation A.1 gives

$$Re = \frac{0.0512 \, \frac{m}{s} \, 0.0381 \, m}{9.03 \times 10^{-7} \, \frac{m^2}{s}} = 2160$$

Other Reynolds number calculations were similar.

---

[1]   J. Kestin, M. Sokolov, and W. A. Wakeham, "Viscosity of liquid water in the range −8 °C to 150 °C," *Journal of Physical and Chemical Reference Data,* vol. 7, pp. 941-948, 1978.

15

CHAPTER 1.10

SPECIAL CONSIDERATIONS FOR WRITING CHEMICAL ENGINEERING REPORTS

As a chemical engineer, communicating the process and results of your experimental work is as important as the work itself. In fact, if you do not share your process and results with the larger community (i.e., your co-workers, your boss, your professional peers), the work itself ceases to have value. A written lab report is one way of documenting and disseminating the details of your research (oral presentations and posters are two other methods). However, just "writing up" an experiment is not adequate; the quality of the writing matters, as does the style of the prose and the format and design of the report. Moreover, your lab report should contain a reasonable interpretation of your results in light of theoretical expectations or literature correlations. If your results do not agree with expectations, then sources of error must be discussed.

Although format and design specifics can vary from organization to organization within the field of chemical engineering, the example report on the following pages – which centers around a fluidized bed experiment—provides you with one common way of constructing, organizing, and formatting a chemical engineering (ChemE) report. The main sections are:

- **Abstract:** Provides a pointed summary of the entire report, focusing mainly on key results and conclusions.
- **Introduction:** Gives the motivation for the study and states the objectives; briefly describes experimental procedure used to achieve objectives.
- **Theory:** Lays out the theoretical concepts that underlie the experiment; explains the method used to analyze the experimental data; details key assumptions.
- **Apparatus and Procedure:** Briefly describes apparatus—usually includes a schematic; References the lab manual from which the procedure was taken and details any significant deviations.
- **Results and Discussion:** Presents and explains the results, using figures and tables to visually display key trends; Discusses results in light of the theory and explains any discrepancies.
- **Conclusions and Recommendations:** Summarizes the key findings and gives an overall "takeway" from the experiment; makes two or three recommendations to improve the lab or suggest the logical next steps in future experimentation.

Audience & chemE lab reports

In telling any story, what you tell, as well as how much context and detail you include, depends on who the listener is. For example, if you are writing an internal memo to your team, you may focus more on the technical details of the project than you would if you were writing an executive summary for the board of directors or the CEO. In writing your lab report, you need to consider such issues. Who would read a lab report? Why? And how much do they already know about the subject?

In general, people would read a report such as the ones you will write because they may find it necessary to undertake a similar study, or they may want to use your findings to help make design or purchase decisions. Given this information, you can assume that your readers have a basic understanding of general chemical engineering principles—for example, fluid, mass, and heat transfer; thermodynamics; reactor design; process control; and so on. In short, any individual with an average knowledge of chemical and/or engineering concepts should be able to read and understand your report without difficulty (Imagine a ChemE or other engineering graduate or an MBA working in a ChBE industry). However, you should recall that even a qualified chemical engineer may have forgotten the specifics of some area of this field. Thus, you may need to remind them of some of the details or clarify the operation of specific units.

Page layout in lab reports

The example Chemical Engineering report that is presented here looks different from the example reports presented elsewhere in this book. This report, for example, uses two-column page layout, it presents a mixture of figure sizes, and it has a lengthy reference list that uses the American Chemical Society (ACS) format style. While these layout requirements are certainly not trivial for the students who must use them in their reports, readers of this book should see these document features as small concessions to audience requirements. Many Chemical Engineering programs ask their students to use the two-column format that is commonly used in professional journals, because this page layout seems to present a more professional appearance for readers. ACS citation and reference style is the default format for students in chemistry and chemical engineering; ACS is, after all, the professional society for professionals in these domains, and it is appropriate for students to learn how to use ACS style in their reports.

The two-column format of this example report constrains the size of routine displays, such as Figures 1, 2, and 5. However, the students' main results, in Figures 3 and 4, are presented using the larger displays that we have discussed elsewhere in this book. It is noteworthy that the small results plot in Figure 5 has been formatted for ease of reading, with axis labels that roughly match the size of the surrounding text.

It is important for readers of this book to see the page design and reference style as small matters of document management. Using an editing program, the document's column layout can be easily changed, and reference management software allows the ACS references to be easily converted to a different format, such as IEEE or APS. What is more important than these page display issues is the information design of this document, as it matches the information design of the other examples presented here. This report begins by motivating the project and stating a specific measurement objective. It describes a method for taking measurements it defines the analytical approach that is taken, and it defines the assumptions that govern the analysis. Finally, it presents results in terms that are specific, that are concrete and that are substantiated by data displays. It provides an Abstract that presents critical numerical results and that draws a conclusion, or "takeaway point." This kind of concrete, specific and brief information presentation is the hallmark of good scientific reporting, regardless of the page layout of the document.

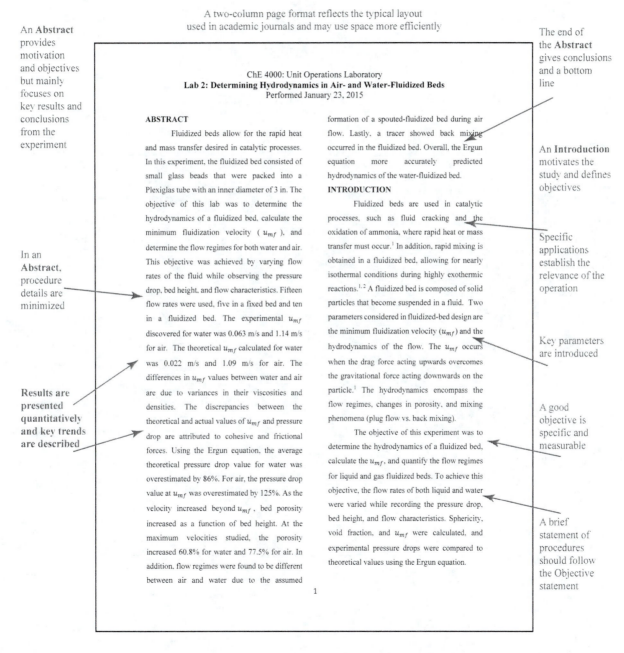

A two-column page format reflects the typical layout used in academic journals and may use space more efficiently

An **Abstract** provides motivation and objectives but mainly focuses on key results and conclusions from the experiment

The end of the **Abstract** gives conclusions and a bottom line

In an **Abstract**, procedure details are minimized

Results are presented quantitatively and key trends are described

An **Introduction** motivates the study and defines objectives

Specific applications establish the relevance of the operation

Key parameters are introduced

A good objective is specific and measurable

A brief statement of procedures should follow the Objective statement

---

ChE 4000: Unit Operations Laboratory
Lab 2: Determining Hydrodynamics in Air- and Water-Fluidized Beds
Performed January 23, 2015

ABSTRACT

Fluidized beds allow for the rapid heat and mass transfer desired in catalytic processes. In this experiment, the fluidized bed consisted of small glass beads that were packed into a Plexiglas tube with an inner diameter of 3 in. The objective of this lab was to determine the hydrodynamics of a fluidized bed, calculate the minimum fluidization velocity (u_{mf}), and determine the flow regimes for both water and air. This objective was achieved by varying flow rates of the fluid while observing the pressure drop, bed height, and flow characteristics. Fifteen flow rates were used, five in a fixed bed and ten in a fluidized bed. The experimental u_{mf} discovered for water was 0.063 m/s and 1.14 m/s for air. The theoretical u_{mf} calculated for water was 0.022 m/s and 1.09 m/s for air. The differences in u_{mf} values between water and air are due to variances in their viscosities and densities. The discrepancies between the theoretical and actual values of u_{mf} and pressure drop are attributed to cohesive and frictional forces. Using the Ergun equation, the average theoretical pressure drop value for water was overestimated by 86%. For air, the pressure drop value at u_{mf} was overestimated by 125%. As the velocity increased beyond u_{mf}, bed porosity increased as a function of bed height. At the maximum velocities studied, the porosity increased 60.8% for water and 77.5% for air. In addition, flow regimes were found to be different between air and water due to the assumed formation of a spouted-fluidized bed during air flow. Lastly, a tracer showed back mixing occurred in the fluidized bed. Overall, the Ergun equation more accurately predicted hydrodynamics of the water-fluidized bed.

INTRODUCTION

Fluidized beds are used in catalytic processes, such as fluid cracking and the oxidation of ammonia, where rapid heat or mass transfer must occur.[1] In addition, rapid mixing is obtained in a fluidized bed, allowing for nearly isothermal conditions during highly exothermic reactions.[1,2] A fluidized bed is composed of solid particles that become suspended in a fluid. Two parameters considered in fluidized-bed design are the minimum fluidization velocity (u_{mf}) and the hydrodynamics of the flow. The u_{mf} occurs when the drag force acting upwards overcomes the gravitational force acting downwards on the particle.[1] The hydrodynamics encompass the flow regimes, changes in porosity, and mixing phenomena (plug flow vs. back mixing).

The objective of this experiment was to determine the hydrodynamics of a fluidized bed, calculate the u_{mf}, and quantify the flow regimes for liquid and gas fluidized beds. To achieve this objective, the flow rates of both liquid and water were varied while recording the pressure drop, bed height, and flow characteristics. Sphericity, void fraction, and u_{mf} were calculated, and experimental pressure drops were compared to theoretical values using the Ergun equation.

1

A **Theory** section explains key concepts relevant to the data analysis

The first paragraph of **Theory** picks up where the **Introduction** left off

Equations are generally centered on the line, with the number right-aligned.

An equation editor such as MathType will give good results.

Equations are treated as part of the preceding sentence and are punctuated accordingly

THEORY

Fluidization occurs when a bed of solid particles is subject to the flow of either a liquid or gas from below.[3] Liquid fluidization, which is more readily explained by the Ergun equation, results in homogenous fluidization, while gas fluidization leads to heterogeneous fluidization. The heterogeneous fluidization consists of two phases, bubbling and particle emulsion, which both occur at velocities slightly above the u_{mf}.[4]

The u_{mf} can be calculated by Equation 1:[5]

$$\left(\frac{d_p^2 \rho_g (\rho_s - \rho_g) g}{\mu^2}\right) = \frac{1.75}{\phi_s \varepsilon_{mf}}\left(\frac{d_p u_{mf} \rho_g}{\mu}\right)^2 + \frac{150(1-\varepsilon_{mf})}{\phi_s^2 \varepsilon_{mf}^3}\left(\frac{d_p u_{mf} \rho_g}{\mu}\right) \tag{1}$$

The sphericity, calculated by Equation 2, determines the frictional losses and is inversely proportional to pressure drop:[6]

$$\phi_s = \frac{\pi^{\frac{1}{3}} \cdot 6 V_p^{\frac{2}{3}}}{S_p} \tag{2}$$

The bed porosity, calculated by Equation 3, is affected by the particle size distribution:

$$\varepsilon = \frac{V_{Voids}}{V_{Total}} = \frac{V_{H_2O}}{V_{Total}} \tag{3}$$

Small particles can occupy interstices between large particles and cause partial fluidization.[7]

At low velocities, the bed is fixed because the gravitational force is greater than the drag force. The pressure drop can be calculated by the Ergun equation,

$$\frac{\Delta P}{L} = 1.75 \frac{(1-\varepsilon_m)}{\varepsilon_m^3}\frac{\rho_g u_o}{\phi_s \overline{d_p}} + 150\frac{(1-\varepsilon_m)^2 \mu}{\varepsilon_m^3}\frac{u_o}{(\phi_s \overline{d_p})^2} \tag{4}$$

Figure 1 shows that the pressure drop for a fixed bed increases linearly with fluid velocity.

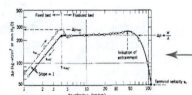

Figure 1. Pressure drop across a packed bed as a function of fluidization velocity.[7]

When the fluid reaches u_{mf}, the greatest pressure drop occurs. After this point, the drag force and gravitational forces are equal, resulting in a fluidized bed. This equality is maintained at higher velocities by an expansion of the bed. The rising height increases the porosity of the bed until the drag and gravitational forces are equivalent.[1]

The fluidization regime depends on the fluidization velocity and can be determined by the Reynolds number (Re), the ratio of inertial forces too viscous forces. The first part of the Ergun equation is the turbulent component (Re<1) while the latter is the laminar component (500<Re<200,000). Relevant Re equations can be found in Appendix A. The viscous forces are dominant at low Re values whereas the inertial forces are dominant at high Re values.[8] All terms in equations are defined in the Nomenclature.

APPARATUS & PROCEDURE

The packed bed used in this experiment included a large number of glass beads enclosed in a Plexiglas tube with an inner diameter of 3 in. A distributor plate was used to minimize deviations in the fluidizing velocity and to stabilize fluidization.[9] Other equipment used

Relevant figures from scientific literature lend credibility

Always credit content obtained from outside sources; this superscript citation uses American Chemical Society style

This reference to the previous sentence helps to make the paragraph flow smoothly

A separate Nomenclature section can be used to define the symbols and terms used in equations

The **Apparatus & Procedure** section describes the equipment and methods used

2

In a short report, procedures may not be described in detail; instead, they may be included in an appendix

included a beaker, electronic scale, thermometer, manometer, air flow meter, water flow meter, syringe, and a dye tracer. Seventeen flow rates of water, ranging from 0.015 m/s to 0.130 m/s, and 15 flow rates for air, ranging from 0.010 m/s to 0.241 m/s, were tested. A schematic of the fluidized bed apparatus is shown in Figure 2 and a more detailed procedure can be found in Appendix B.

A diagram should represent the experimental setup.

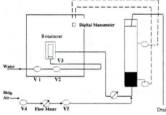

Figure 2. Fluidized bed apparatus.

Hand-drawn diagrams are seldom adequate, but simple drawing tools often give good results

The experimental procedure was taken from the UO Lab Manual[5], without any deviations. The safety procedures followed during this experiment included never turning the water and air valves on simultaneously and slowly turning on the flows, to ensure that beads did not escape the column. Standard safety protocol, including the use of 100% cotton lab coats, closed-toe shoes, safety glasses, and gloves, was followed.

Results & Discussion sections report key findings and analyze their significance

RESULTS & DISCUSSION

The purpose of this experiment was to explore the fluidized-bed hydrodynamic effects, determine the u_{mf}, and assess the flow regimes

The project goals are first reviewed, to establish context for the findings

of fluidization. The hydrodynamic effects were quantified with the calculation of parameters such as the particle sphericity, the fixed bed porosity, and the u_{mf}. Assuming oblate spheroids, a particle sphericity of 0.9998 was found. A sphericity of one would be a perfect sphere. A fixed bed porosity of 0.34 was also determined. Calculations are provided in Appendix C.

The u_{mf} values of air and water were found experimentally by varying the fluid velocities and theoretically by using the Ergun equation. Figure 3 below shows that the experimental u_{mf} for water was 0.063 m/s, whereas the theoretical u_{mf} for water was 0.022 m/s. The experimental u_{mf} correlates to the onset of the plateau shown in Figure 3. According to this figure, the pressure drops determined by the Ergun equation exceed the experimental values and underestimates the u_{mf}. The average theoretical pressure drop value for water was overestimated by 86%. The discrepancies in the theoretical and experimental values can be attributed to cohesive forces, frictional forces, or a combination of the two. Cohesive forces exist between particles and between the particles and the distributor plate. Frictional forces exist at the walls of the vertical column.[10] Additionally, discrepancies could be due to human error and systematic error from the calibration of the rotameter that caused fluctuations in the pressure drop values.

Each figure must be introduced before it appears. If the figure appears on a following page, use the word "below"

Discuss and interpret the key points of any figure you present

Sources of error should also be explained in the Results section

3

You can break the two-column format to display large figures, so long as these are important. Use section breaks to cleanly change format schemes

Data sets are distinguished by different markers and colors

Scatter plots are useful for showing trends in data and comparing data to predictions

The display is enhanced when the Legend is placed on an empty area of the plot

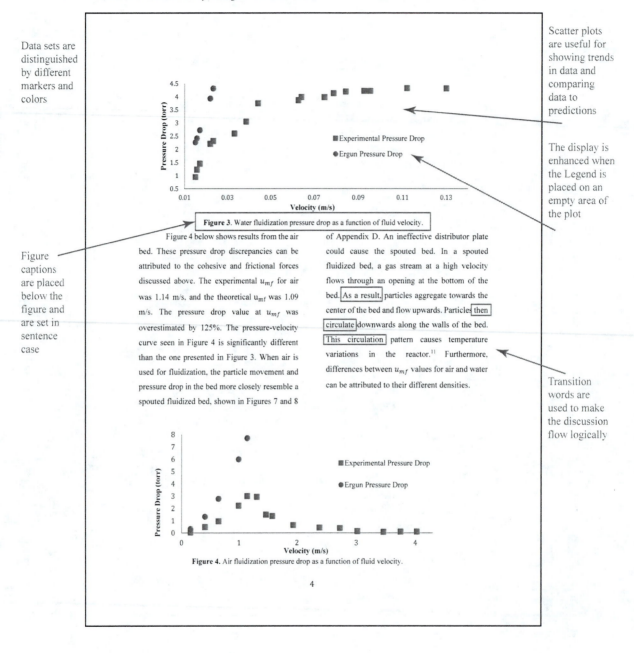

Figure 3. Water fluidization pressure drop as a function of fluid velocity.

Figure captions are placed below the figure and are set in sentence case

Figure 4 below shows results from the air bed. These pressure drop discrepancies can be attributed to the cohesive and frictional forces discussed above. The experimental u_{mf} for air was 1.14 m/s, and the theoretical u_{mf} was 1.09 m/s. The pressure drop value at u_{mf} was overestimated by 125%. The pressure-velocity curve seen in Figure 4 is significantly different than the one presented in Figure 3. When air is used for fluidization, the particle movement and pressure drop in the bed more closely resemble a spouted fluidized bed, shown in Figures 7 and 8

of Appendix D. An ineffective distributor plate could cause the spouted bed. In a spouted fluidized bed, a gas stream at a high velocity flows through an opening at the bottom of the bed. As a result, particles aggregate towards the center of the bed and flow upwards. Particles then circulate downwards along the walls of the bed. This circulation pattern causes temperature variations in the reactor.[11] Furthermore, differences between u_{mf} values for air and water can be attributed to their different densities.

Transition words are used to make the discussion flow logically

Figure 4. Air fluidization pressure drop as a function of fluid velocity.

4

A Topic Sentence should begin a paragraph

Reynolds numbers (Re) were also found for both air and water. Water maintained Re values between laminar and turbulent flow. However, air transitioned into fully turbulent flow at high velocities. See Appendix C for data.

In addition to determining the u_{mf}, the residence time of 1 mL of tracer dye was found so as to characterize the hydrodynamics of the fluidized bed. The observed experimental residence time was 1.15 sec, while the theoretical plug-flow value was determined to be 0.74 sec. Plug flow reactors ideally have no back mixing or dead zones. For perfectly back-mixed vessels, the residence time should be prolonged and approach infinity.[12,13] The experimental value therefore indicates that the fluidized bed is neither a perfectly plugged flow nor back-mixing vessel.

A concluding sentence can emphasize the paragraph's point, particularly if the point is complex

The last hydrodynamic property investigated was the change in porosity. The bed expanded as the velocity increased. Since porosity depends on height, as the height increased the porosity increased. The relationship between porosity and velocity can be seen in Figure 5.

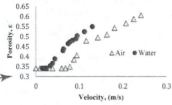

Figure 5. Porosity as a function of velocity.

Simple figures can be placed in a column, so long as axes, legends and markers are legible

Porosity calculations and equations can be found in Appendix C. In order to see if there is a difference between the ε values for water and air, a t-test was completed. This t-test showed with a 90% confidence that the ε values were not significantly different.

Calculations and derivations may be placed in an appendix, but they must be called out in the report text

CONCLUSIONS & RECOMMENDATIONS

This experiment was successful in showing that fluidization is directly influenced by particle properties (sphericity, diameter, and density), fluid properties (velocity, viscosity, and density), and bed properties (porosity, height, inner diameter, and distributer plate). It was discovered that the experimental and theoretical u_{mf} values were different for both air and water. These differences were due to cohesive and frictional forces in addition to systematic error and random noise. Another discovery was that the air bed followed a spouted-fluidized-bed model more accurately than a homogenous fluidized-bed model. Spouted-fluidized beds result in non-uniform distribution of temperatures and therefore would be undesirable for isothermal catalytic processes.

For the future, we recommend improving this experiment by testing the distributer plate to determine if non-uniform distribution is occurring. Another recommendation would be to find a better correlation that corrects for the discrepancies between theoretical and experimental u_{mf} values caused by cohesive and frictional forces.

Conclusions & Recommendations sections should summarize key points and may suggest future improvements

Conclusions should first define success or failure of the project

Conclusions should end with a "bottom line" statement. Why should a reader care about your results?

Recommendations may address error sources or safety issues, or they may suggest new experimental approaches

5

This reference list is formatted to the American Chemical Society style guide, which is commonly used in Chemical Engineering courses

REFERENCES

1) Fluidized-Bed Reactors, University of Michigan Professional Reference Shelf Chapter 12 3rd Edition. http://www. umich.edu/~elements/12chap/html/Fluid izedBed.pdf (accessed February 2nd, 2014).

2) Subramanian, Shankar R.; Flow through Packed Beds and Fluidized Beds, Clarkson Projects Web site. http://web2. clarkson.edu/projects/subramanian/ch30 1/notes/packfluidbed.pdf (accessed February 2nd, 2014).

3) Hesketh, Robert P. Fluidization. http://users.rowan.edu /~hesketh/0906-309/Laboratories/Fluidized%20Bed%20 flow%20regimes%20rev1.pdf(accessed February 3, 2014).

4) Thermopedia: Fluidized Bed. http://www.thermopedia.com/content/46 /?tid=104&sn=1297 (accessed February 2, 2014).

5) Unit Operations Lab Manual: Fluidized Bed Lab; Georgia Institute of Technology: Atlanta, Ga. P.1

6) Lee, S; Henthorn, K. Particle Technology and Applications (pg. 150). http://books.google.com/books (accessed February 3, 2014).

7) Fluidized Beds. http://www. umich.edu/~elements/12chap/html/Fluid izedBed.pdf (accessed February 2, 2014).

8) Thornhill, Doug. Fluid Mechanics for Fluidized Beds. http://faculty. washington.edu/finlayso/Fluidized_Bed/ FBR_Fluid_Mech/fluid_bed_scroll.htm (accessed February 2, 2014).

9) Airflow in Batch Fluid Bed Processing. GEA Process Engineering Inc. http://www.niroinc .com/pharma_systems/airflow_batch_flu id_bed.asp (accessed Febrary 5, 2014).

10) Srivastava, A; Sundaresan, S. Role of wall friction in fluidization and standpipe flow. https://www.princeton.edu/cbe/ people/faculty/sundaresan/group/publica tions/pdf/90.pdf (accessed February 3, 2014).

11) Sutkar, V; Deen, N; Kuipers, J. Spout fluidized beds: Recent advances in experimental and numerical studies. *J.ELSEVIER. A* [Online] **2013**. *86,* 124-136. www.elsevier.com (accessed February 3, 2014)

12) Fogler, S; Elements of Chemical Reaction Engineering, 4th ed.; Prentice Hall; NJ, USA pg. 916

13) Burbidge , A.; Georget, E.; Mathys, A.; Sauvageat, J.; Residence time distributions in a modular micro reaction system. *Journal of Food Engineering.* **2013**, *Volume* 116 (4), 910-919.

14) Weisstein, Eric W. "Oblate Spheroid." From *MathWorld*--A Wolfram Web Resource. http://mathworld.wolfram. com/OblateSpheroid.html

15) Geankoplis, Christie. Transport Processes: Momentum, Heat and Mass (pg. 125). http://my.Safaribooksonline. com/book/chemical-engineering/01310 1367x/transport-processes-momentum-heat-and-mass/part01 (accessed Febuaury 4, 2014).

16) Airflow in Batch Fluid Bed Processing. GEA Process Engineering Inc. http://www.niroinc.com/pharma_system s/airflow_batch_fluid_bed.asp (accessed Febrary 5, 2014).

6

Chapter 1.11

Example of an Undergraduate Research Report

Undergraduates and Research

One of the more mysterious aspects of university life for undergraduate students is what professors mean when they mention their "research." Undergraduate students know professors and teaching assistants are somehow involved in "research," but often they do not know what that research is, how it is conducted, or how they can become involved in it.

Professors typically spend 40% of their time teaching, 40% on research, and 20% on administrative duties. So what is research? It is really the search for knowledge and the attempt to understand how things work. For many undergraduate projects, your research consists of finding and reviewing things that have already been discovered or are well-known. The research conducted by professors and graduate students examines things that are unknown or poorly understood. We do this by studying, testing, and analysing things (or maybe aspects of things) that have not been studied before. We may use laboratory experiments, field measurement or monitoring, or complex computer models. For engineering researchers, we are also interested in how this knowledge can be applied to real world designs.

Researchers will be interested in a very broad area (e.g., as a structural engineering researcher, I might be interested in how bridges react when certain loads are applied), but their work will involve a very small and specialized aspect of this broader topic (e.g., How do steel bridges resist repeated loads that result in metal fatigue?). It can be intimidating when you first encounter a research project, because you will find that there seems to be a mysterious language spoken by those working in that sub-speciality! These will be the detailed terms and concepts that are used to define the physical models that researchers study. Although overwhelming at first, you will find that with continued practice and learning, this "language" will start to make sense, and you will be able to converse with others in this sub-specialty.

Professors typically work closely with graduate students on research projects. Professors will identify and define research topics, and obtain funding for the research. Graduate students (either master's or doctoral) will work with the professor to define the methodology for answering the research question, will supervise or conduct the experiments, and then analyse the data or results obtained. Together with their professor they will attempt to understand what the results mean and the impact on the wider engineering community.

The results are usually published as "papers" which summarize the important results from the work. The graduate students and professor will work together on drafts of the paper until it is ready to submit to a journal or conference. Journals are the publications in which research findings are disseminated. A journal will usually have a very specialized focus and will only accept papers describing research in that area. Once a paper is submitted, it is sent out for "review." When a paper is under review it means the paper will be read and critiqued by a number of experts in that field. The reviewers carefully examine the methodology, results, and analysis. If they detect any errors in the paper (e.g., a methodology that has been incorrectly applied, or mistakes in the analysis that leads to incorrect conclusions), the paper will be rejected. Another reason for rejecting a paper is if the work is deemed to simply be a copy of research that has already been published. If the reviewers do not find such errors, they will recommend that the paper be published. Note, however, that even though these reviewers did not find errors, others may find errors in the future! In addition, future research may show that some conclusions from earlier research are wrong or incomplete.

To better understand the research going on in your department, a good place to start is the department website. Usually the department will list research interests and recent projects along with the biography of each professor. If there is a broad area you are interested in (e.g., maybe as a Civil Engineering student, you are interested in structures) focus on the professors who work in that area, and then look at the sub-specialities of each of them.

So is there a place for undergraduate students in research? Yes! There are a number of ways an undergraduate might get involved. Some professors will hire undergraduate students (usually those completing their 2nd or 3rd year) as "research assistants". The research assistant will often work closely with a graduate student to help install instrumentation, set up experiments, conduct simple analysis or literature review, or similar tasks. At this stage, you are really learning the "language" of the sub-discipline and of research itself.

These sorts of jobs may be advertised by the department, or you can contact your department faculty advisor or the Undergraduate Chair. These are professors who will be familiar with the research in the department and can let you know about research opportunities for undergraduates. If there is an area of research you are interested in, you can try to directly contact (usually by e-mail initially) the professor that works in that area. There are limited numbers of these jobs, so it can be quite competitive!

Many engineering departments also have a senior year thesis course. Note that this is usually different than the capstone design course most senior engineering undergraduate students must take. In many cases, the thesis course takes the form of a small research project supervised by a professor and led by a 4th year student. Obviously, because an undergraduate student has much less time to devote to the research than a graduate student, the scope and focus will usually be reduced compared to graduate research. At the end, you will write and present a research paper or thesis, which will take a form similar to a journal paper, albeit a much shorter and less detailed version. An undergraduate research thesis is an excellent way to discover whether graduate school and research might be for you.

If you discover you have an interest and passion for research, you might consider graduate school. A good way to start is by contacting, early in your final year, the Graduate Chair for your department. This is a professor who will be familiar with the application process, including deadlines and the necessary paperwork.

Research Papers

In this section, a paper written on the research conducted by an undergraduate student will be discussed. This paper was a "conference paper," meaning that it is a short paper that is included as part of the conference proceedings. Conference papers are typically shorter (about 10 pages) and less-detailed than journal papers. However, they will be reviewed by experts in the field before they are accepted (or rejected) and published. Many undergraduate research projects are ideal for conference papers.

Although this paper is shorter than a journal paper, its outline and presentation is very similar to that expected for a journal paper. This example will show some of the differences between the presentation of a research paper and an undergraduate lab report or a capstone design project report.

EFFECT OF CEMENT, LIME, AND BIORESIN STABILIZERS ON COMPRESSED EARTH BLOCK PERFORMANCE

Kenneth Mak[a], Oke Maracle[b], and Colin MacDougall[c*]

[a]PhD Candidate, Department of Civil Engineering, Queen's University, Canada; kennethdmak@gmail.com

[b] Research Associate, Department of Civil Engineering, Queen's University, Canada; 11om3@queensu.ca

[c*]Associate Professor, Department of Civil Engineering, Queen's University, Canada; macdougc@queensu.ca

*Corresponding author

Keywords are included to make it easier for those searching on databases to find your research.

Keywords: Compressed earth blocks, Portland cement, bioresins, freeze, moisture absorption.

Abstract: Earth construction has been traditionally used in warm, arid climates; however, in recent years, there has been growing interest in its application in cold climates, like Canada. Stabilized compressed earth blocks (CEBs) have been used in a number of residential applications, but the long-term performance and moisture susceptibility of CEBs remain a concern. In this paper, a series of new stabilizers, including Permazyme and bio-resins, have been tested and compared to conventional stabilizers for their compressive performance. To be able to determine the impact of moisture on durability, moisture absorption rates were measured. Fully saturated CEBs with these stabilizers then were subjected to -10 °C temperatures until fully frozen. Their compressive strengths were compared to dry and fully saturated CEBs tested at room temperature. Notably, CEBs containing 5% bio-resin demonstrated a 61% to 109% increase in dry compressive strength compared to 5% cement blocks, whereas Permazyme showed no significant increase in dry capacity compared to unstabilized blocks. These results demonstrate that Permazyne (usually used to improve moisture resistance) cannot be used as a substitute for cement as a CEB stabilizer, but bio-resins show potential as a CEB stabilizer.

An abstract provides a summary (typically 300 words or less) of the paper, including a brief description of the problem, methodology, and key conclusions.

Note that although the Keywords and Abstract appear at the start of the paper, they are often the last elements written!

3

The Introduction serves a number of key purposes. First, what is it you are researching? In this paper, the first words tell you it is "Earth-based construction...in Canada."

Second, the Introduction sets the overall context for the research. Why is this research important? In this paper, the first paragraph provides some of the motivation for the research ("Earth-based construction is relatively new..."; "...builders encounter skepticism...").

Introduction

Earth-based construction is relatively new in Canada. The builders involved in this niche market tend to be small, and the projects are usually custom-built homes. These builders encounter skepticism from building officials in terms of the durability of earth-based construction. Thus, they usually take a conservative approach and use high levels of stabilizers like Portland cement. However, they recognize that from an environmental perspective, Portland cement is not optimal.

This paper describes testing conducted at Queen's University in collaboration with a small local builder. This builder was interested in finding alternatives to Portland cement for a stabilizer for the compressed earth blocks (CEBs) he manufactures. He was also interested in obtaining objective information on the moisture resistance of the CEBs and the strength in sub-zero temperatures. Therefore, an investigation on the performance of compressed earth blocks, separated into two phases, was conducted. Phase 1 focuses on the potential application of new stabilizers for use in compressed earth blocks. The study consisted of compressive testing of a limited number of lab-manufactured blocks. Phase 2 focuses on the impact of moisture on compressed earth blocks. The study consisted of moisture absorption tests and preliminary freeze tests.

The Introduction will also explain how your work fits in with research that has been conducted before. This might take the form of a brief Literature Review, with references provided for previous research. In our case, there had been little previous testing in this specific area, so there was no literature to review!

Finally, the Introduction provides a "road map" for the rest of the paper. There will be some key questions you want to investigate. These need to be clearly stated for the reader. For this paper, the research is divided into two "phases," which are really the two overall questions we were investigating.

> The methodology of the research is described next. You must provide enough relevant detail that someone who wanted to repeat your work could do so. This paper reports experimental work, so it was important that we clearly explained the test methods used, the materials, how test specimens were prepared, and descriptions of any test apparatus. Obviously, the type of information you must provide depends on your methodology. However, all of it needs to be reported in a very concise way to stay within page limits!

Experimental program

This section details the test specimens and parameters; materials; fabrication of specimens; and test setup and instrumentation.

Test specimens and parameters: Portland cement is the most common sta[...]
Phase 1, five alternative stabilizers were investigated to quantify their impa[...]
strength. In Phase 2, CEBs were tested using three distinct approaches to q[...]
and water-related performance. The first set of tests (*2R*) investigated the a[...]
applied to the CEB to resist wind driven rain. RILEM Test Method II.4 [1] [...]
rain on the surface of the CEBs. The second set of tests (*2A*) investigated th[...]
of CEBs based on ASTM C140 [2]. The final set of tests (*2F*) investigated the strength of fully saturated CEBs that are subsequently subjected to sub-zero temperatures typical of Canadian winters.

> If you use "test standards" (e.g., an ASTM test standard), provide a reference to that standard, as it will save having to write out many details of the test method!

Three repetitions were done for each specimen type unless otherwise stated. Individual block types and properties are shown in Tables 1 and 2.

Materials: Compressed earth blocks were manufactured using the following materials:

Soil: Site soil from Coburg, Ontario, Canada was used for the compressed earth blocks. Tests by a commercial testing lab [3], including x-ray diffraction and x-ray fluorescence analysis, indicate that the soil contains 23.3% by mass of clay-size material. X-ray diffraction indicated that the majority of the clay-sized material was calcite.

Cement: Type I Portland cement is a commonly used stabilizer for CEBs.

Lime: Hydrated dolomite lime is also commonly used to stabilize CEBs [4].

Metal Oxide: Metal oxide is used to modify the colour of CEBs.

Permazyme: A natural enzyme mixture, which is commonly used in the process of oil stabilization, was used as a CEB stabilizer [5].

> You must provide enough detail that readers can identify exactly what materials were used in your research. A reference to a manufacturer's web-site is one way to do this.

Tables are a very efficient way to show the reader the full scope of specimens being tested for an experimental program. For other research methodologies, you might use a table to show the various parameters being investigated.

Table 1: CEB Types

Specimen Type	Stabilizer							Water Repellent Concentration [%]
	Cement [% of soil weight]	Lime [% of soil weight]	Metal Oxide [% of soil weight]	Bioresin [% of soil weight]	Epoxy [% of soil weight]	Nut Epoxy [% of soil weight]	Permazyme [g/yd^3]	
1C-1	5							
1C-2			5					
1C-3				5.15				
1C-4					6.73			
1C-5						7		
1C-6							1	
2R-1	5							0
2R-2	5							5
2R-3	5							6.7
2R-4	5							10
2R-5	5							25
2A-1	5							
2A-2	7.5							
2A-3	10							
2A-4	5	2.5						
2F-1	5							
2F-2	7.5							

Bioresin: A bioresin, comprised of a commercial available furfuryl alcohol ($C_5H_6O_2$) was used as a CEB stabilizer. The bioresin was derived from renewable resources, such as sugarcane and corncobs. It was hardened with a commercially available catalyst at a ratio of 3% catalyst-by-weight. The catalyst consisted of phthaloyl chloride ($C_8H_4O_4Cl_2$) and dimethyl phthalate ($C_{10}H_{10}O_4$).

Here we have provided each type of test specimen with an identifying code. Later when we present results, the reader can reference this table to quickly identify the type of specimen being discussed.

Epoxy: A commerc[...] reported post-cured (curing at 60°C for 72 hrs) tensile strength an[...]8.18 GPa, respectively. It exhibits a maximum elongation of 5% [...]

Nut Epoxy: A com[...]n conjunction with an alkylated phenolic polyamine as its ha[...]ed from cashew nut shell oil and used at a weight ratio of 100 parts r[...]

Water Sealant: A commercially available blend of silanes and oligomeric alkoxysiloxanes, which is commonly used as a penetrating water repellent for dense or porous masonry surfaces. The mixture was provided in a concentrated form and diluted to specified concentrations [7].

Table 2: CEB Properties

Specimen	Phase	Test Type	Dimensions			Density [kg/m³]
			Length [mm]	Width [mm]	Height [mm]	
1C-1	1	Compression	50.0	50.0	31.4	2176.9
1C-2					33.0	2207.7
1C-3					31.5	2243.5
1C-4					30.6	2244.8
1C-5					30.4	2214.0
1C-6					31.7	2182.0
2R-1	2	RILEM	177.8	355.7	93.8	2033.4
2R-2						
2R-3						
2R-4						
2R-5						
2A-1		ASTM	177.9	355.7	93.1	2094.8
2A-2			177.4	355.7	94.2	2122.9
2A-3			177.2	355.8	93.0	2142.0
2A-4			177.8	355.8	93.9	2079.3
2F-1		Freeze	177.8	355.6	279.6	2094.8
2F-2			177.8	355.6	273.0	2122.9

Fabrication of specimens: Specimen fabrication method varied based on the test phase, as described below:

Phase 1: CEBs were manufactured in the lab. Site soil was delivered and used as the primary ingredient with additives, as outlined in Table 1.

Due to the preliminary nature of this phase, 50 mm x 50 mm x 50 mm cube samples were prepared. Site soil, additives and water were manually mixed and placed in a steel mould. A steel bearing pad was placed between the soil and the loading head of a Lab Integration universal testing machine. The surface was loaded at a rate of 6 mm/min up to a pressure of 10 MPa to replicate the loading pressure of typical site-manufactured CEBs. Specimens were demoulded immediately and left uncovered to cure for a minimum of 30 days prior to testing. The final height of each specimen is indicated in Table 2.

Phase 2: CEBs were manufactured by an experienced builder in Cobourg, Ontario, Canada. Site soil was used as the primary ingredient with additives, as outlined in Table 1.

An AECT 3500 Compressed Earth Block Machine was used to manufacture CEBs of approximate width 356 mm and length 178 mm. The final height of each specimen is indicated in Table 2. Soil and additives were combined in a mixer, and then water was added until the desired consistency. An average compressive stress of 10.7 MPa was applied by the machine to fabricate the blocks. All blocks were cured for a minimum of 30 days prior to transportation to the lab.

Specific specimens were treated with a water sealant. The water sealant concentrate was diluted with water as per the ratio defined in Table 1. As per manufacturer's instructions, the solution was applied uniformly with a roller and the surface was maintained wet for 3 minutes. Excess was wiped off. Specimens were allowed to dry for a minimum of 10 days prior to testing.

> After describing the materials we used, we then had to describe how the materials were put together to form the samples we ultimately tested. Depending on your research methodology, this may or may not be relevant in your paper.

Test setup and instrumentation: *Phase 1*: Compressive tests of specimens were performed on a universal testing machine. Dimensions and mass of every block were measured and recorded prior to conditioning as well as before testing. CEBs were loaded between 0.05 MPa/s and 0.20 MPa/s using spherically-seated steel plates, such that failure would be induced within 1-2 minutes from 50% ultimate load as per ASTM C1314 [8]. Sheets of 3 mm thick plywood were used to cap the specimens. Maximum load and original cross-sectional area were used to determine the strength of each block.

Phase 2: Water absorption was tested utilizing two different standards. RILEM Test Method II.4 [1] defines water absorption based on an uptake tube test method. A short RILEM tube was affixed to the side of the CEB with putty. Gradations were present on the side of the tube, with each gradation representing a specific head pressure and in effect, an equivalent wind-driven rain. Typically, water is added to a specified gradation to determine the water absorption relevant to the specified conditions. For this series of tests, water was added to reach the highest graded line. Measurements of the gradation level were made at specified periods. The RILEM test was performed on all sides of the specimen.

ASTM approaches water absorption and its impact two different ways. Firstly, ASTM D559 [9] is targeted towards earthen materials and assumes water absorption to occur via capillary action on the bottom of the specimen. Secondly, ASTM C140 [2] is targeted towards masonry units and determines water absorption based on submersion. It was observed that over several days of exposure, which is normal in practice, both methods yield the same saturation level with the latter being faster [10]. The second approach was adopted, whereby specimens were fully submerged in water. There was a minimum of 25 mm of water from the CEB surface to the surface of the water. Specimens were removed, lightly wiped to remove surface puddles and weighed at specified intervals.

Freeze testing of specimens consisted of compressive testing of frozen, fully saturated CEBs. Firstly, CEBs were submerged for 24 hours to reach a maximum level of saturation, as determined from the water absorption tests. Secondly, CEBs were exposed to exterior temperatures of at least below -10°C for a 24 hour period. Additional CEBs were implanted with thermocouples and exposed to the same conditions to ensure the internal temperature was below 0°C. Lastly, compressive testing of the frozen CEBs was performed using a Forney testing machine. Internal and external temperature measurements were taken to ensure that the CEB was fully frozen upon testing.

Dimensions and mass of every block were measured and recorded prior to conditioning as well as before testing. CEBs were loaded between 0.05 MPa/s and 0.20 MPa/s using spherically-seated steel plates, such that failure would be induced within 1-2 minutes from 50% ultimate load as per ASTM C1314 [8]. Sheets of 3 mm thick plywood were used to cap the specimens. Maximum load and original cross-sectional area were used to determine the strength of each block.

"Results" refers to the presentation of the results of the experiments. "Discussion" interprets and analyzes the results, and attempts to identify the key parameters that govern the process we are studying. In a journal paper, these are often separate sections, while in conference papers they may be combined. as they are here.

Experimental results and discussion

This section presents the results of the experimental program. This includes the effect of additives on compressive strength, effect of a water sealant on absorption rate, effect of common additives on absorption rate, and effect of moisture on mechanical performance when frozen.

Due to the variation of aspect ratios amongst tests, a correction factor was implemented to provide an equivalent unconfined strength. Heathcote and Jankulovski's correction factor [11] was selected, as per suggestion by Mak et al. [12].

Effect of additives on compressive strength: Six CEB types were tested to determine the impact of additives. Results are shown in Table 3. CEB 1C-1, which contains 5% cement, is the most common CEB mix for construction. As such, it represents the baseline at 3.85±0.15 MPa. This baseline is representative of this manufacturing method.

Here a table is used to present the results of a number of compressive strength tests.

Table 3: Phase 1 Compressive Testing Results

Specimen	Repetitions	Results		Heathcote and Jankulovski Correction			
		Strength [MPa]	Std. Dev. [MPa]	Aspect ratio [h/t]	Correction Factor	Strength [MPa]	Std. Dev. [MPa]
1C-1	3	10.60	0.41	0.63	0.36	3.85	0.15
1C-2	3	3.95	0.72	0.66	0.38	1.50	0.27
1C-3	3	17.11	1.44	0.63	0.37	6.25	0.53
1C-4	3	34.57	2.62	0.61	0.36	12.32	0.93
1C-5	3	22.18	2.38	0.61	0.35	7.84	0.84
1C-6	3	5.01	0.46	0.63	0.37	1.84	0.17

CEB 1C-2, which used metal oxide as the stabilizer, had a compressive strength of less than 50% that of CEB 1C-1. Metal oxide is currently being added to the CEBs for aesthetics, and clearly cannot be used as a direct replacement for cement. A similar capacity was observed in 1C-6, which used only permazyme as a stabilizer. Permazyme is primarily intended to improve erosion resistance. Again, the results indicate it cannot be used as a direct replacement for cement to obtain a target compressive strength.

Phrases like "...may be attributed to..." show that an interpretation of the results is being made. Sentences like this are the "discussion" part of this section.

The use of resin as a stabilizer f[...] promising in terms of their impact on c[...] was observed when compared to CEB 1C-1. The bioresin-stabilized CEB, 1C-3, showed an increase of 52%, whereas the commercial epoxy-stabilized CEB, 1C-4 resulted in an increase of 220%. CEB 1C-5, which contains the nut epoxy, yielded an increase of 104%. Higher variability compared to the other specimens was observed with the use of resin systems. This may be attributed to dispersion of small quantities of resins.

Despite the promising results, the builder involved with the project was concerned with the cost and environmental impact of resins. Therefore, Phase 2 focused on CEBs with more conventional stabilizers.

Effect of a water sealant on absorption rate: CEBs were surfaced treated with four different concentrations of a water repellent mixture. Untreated CEBs, 2R-1, were used as a baseline. The effect of the treatment is shown for both the 177.8 mm by 93.8 mm side (Figure 1(a)) and the 177.8 mm by 355.7 mm bottom face (Figure 1(b)) of the CEB, where the highest graded line represents the initial starting point. No movement from the "highest graded line" represents a negligible amount of water absorption. A minimum of three repetitions were performed between the two faces. All points have been plotted.

In all instances, the fastest water absorption rates were observed for the CEBs that were not

> Figures are one of the key ways we present data, illustrate trends, and make comparisons. In writing a paper, most of your time should be spent on planning, producing, and refining your figures! They are what truly tells the story of your research. Every figure must be referenced and discussed in the text. It is not enough to simply refer the reader to the figure. What are the keys points you want them to see?

repellent; however, it is not a direct correlation between water absorption and water repellent. This may be caused by clay platelet swelling at the ingress point.

A clear distinction can be drawn between the faces of the CEB and its impact on absorption rate; however, it is purely qualitative. As shown in Figure 1, the 177.8 mm by 355.7 mm bottom face showed a significantly higher absorption rate than compared to that of the narrower (177.8 x 93.8 mm) side, where four of five CEB types reacted the lowest resistance rating on the RILEM tube within 20 minutes. Therefore, water absorption may be governed by capillary action as opposed to rain-driven water.

> Modelling, theoretical predictions, or other limits may be shown as lines.

> For experimental data, individual test results should be shown as "markers".

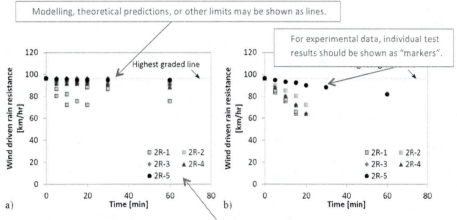

a)   b)

94   Figure 1: Wind driven rain resistance as determined with the aid of the short RILEM tube, with placement of the tube on the (a) 177.8 mm by 93.8 mm side face; and, (b) 177.8 mm by 355.7 mm bottom face.

> Several data sets may be presented in the same figure to permit comparison. Use different markers to clearly distinguish the different

A good practice in presenting any results is to first refer to the relevant figure, then discuss the key points the figure shows:

Effect of common additives on absorption rate: The rate of water absorption, as per ASTM C140 [2], was measured for CEBs with a typical range of cement and lime stabilizer. The results are shown in Figure 2. Three tests were conducted for each stabilizer ratio. A key observation is that the water absorption rates are relatively equivalent regardless of the ratio of stabilizer in the CEBs. Each block absorbs a maximum of approximately 1 kg water. These blocks have an average mass of 12 kg, thus they can absorb approximately 8% of their dry mass. The initial rate of water absorption is very rapid. Within about 25 minutes, the CEBs can absorb 0.8 kg water (80% of their maximum).

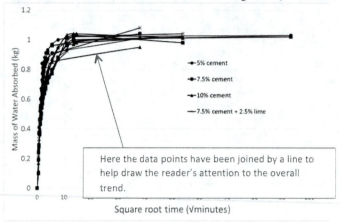

Here the data points have been joined by a line to help draw the reader's attention to the overall trend.

Figure 2: Water absorption rate for CEBs with varying cement and lime stabilizer.

> If your data is presented in a table, again start by referring to the table, then discussing key observations:

Effect of moisture on mechanical performance when frozen: Two CEB types were tested to determine the impact of sub-zero temperatures on compressive strength: CEB 2F-1, containing 5% cement, and CEB 2F-2, containing 7.5% cement. CEBs used for this study were from the same batch as those used in another study, and as such, were compared to the room-temperature dry and wet strengths obtained by Mak et al. [12]. In this instance, the "dry state" occurs when the CEB's moisture content is allowed to reach equilibrium at room temperature. It was determined that there was no significant difference between room temperature dry and oven dry under the specified lab conditions. The "wet state" occurs when a CEB has been submerged in water for 24 hours, which exceeds the time required for full saturation. Results from all of these tests are presented in Table 4.

It is well known that the capacity of CEBs diminishes greatly with increasing water content. This continues to be the case, as shown in Table 4; however, there was a drastic increase in compressive strength when frozen. At 5% cement content, CEBs demonstrated a strength increase of 41%. At 7.5% cement content, CEBs demonstrated a strength increase of 24%. Two phenomena were observed: increase in capacity when frozen, and diminishing strength gain with increasing cement content. The increase in capacity when frozen may be due to the ice acting as a binder for the entire system, similar to addition of cement. Furthermore, the impact of the ice as a binder may be relative to the amount of other binders. With an increase in cement content, there was a decrease in strength gained due to the ice formation.

> Give some thought to the best way to present your data. The test results shown below needed to be "corrected" using an equation from the literature. The easiest way to show the "raw" and "corrected" data was with a table.

Table 4: Phase 2 Freeze Testing Results, with dry (D) and wet (W) test results from Mak et al. [] denoted by *M*- (i.e. M-1D), where 1 represents the equivalent additive mixture to 2F-1 and 2 represents the equivalent additive mixture to 2F-2

Specimen	State	Repetitions	Results		Heathcote and Jankulovski Correction			
			Strength [MPa]	Std. Dev. [MPa]	Aspect ratio [h/t]	Correction Factor	Strength [MPa]	Std. Dev. [MPa]
2F-1	Frozen	3	15.01	0.99	1.57	0.67	10.08	0.66
2F-2		3	17.56	1.08	1.54	0.67	11.69	0.72
M-1D	Dry	3	12.16	0.81	1.02	0.59	7.17	0.48
M-2D		3	16.1	0.31	1.04	0.59	9.46	0.17
M-1F	Wet	3	5.14	0.08	1.02	0.59	3.02	0.04
M-2F		3	7.65	0.71	1.04	0.59	4.83	0.4

The final section is called "Conclusions." The conclusions should be very specific, quantified observations or results that flow directly from your research. There should be no new information presented in this section that has not been discussed previously somewhere in the text of the paper!

Conclusions

This experimental program invested the impact of alternative stabilizers and w properties of various conventional CEB types. The study focused on the effect compressive strength, the effect of a water sealant on water absorption, the eff additives on absorption rate and the effect of moisture on mechanical propertie following conclusions were drawn:

Note that the "increase in capacity" has been quantified by providing the range of percent increase observed.

1. The replacement of cement with metal oxide and permazyme showed no improvement in compressive strength, whereas the replacement of cement with resin systems showed an increase in capacity ranging from 52% to 220%.
2. The use of a water sealant was effective for reducing water absorption in CEBs; however, due to the test method, results were only qualitative. The bottom face of CEBs demonstrated a faster absorption rate.
3. RILEM testing may only provide a rough estimate of water absorptivity of CEBs. This may be associated with the clay platelets and the small contact surface associated with the RILEM tube.
4. All CEBs that were manufactured using conventional stabilizers showed no difference in absorption properties regardless of stabilization.
5. Although moisture content typically decreases CEB compressive strength, it increases it when frozen. This may be associated with the ice acting as a binder.

Sometimes the research methodology only permits a qualitative conclusion, but this is clearly stated.

> Funding agencies or industry partners are typically thanked in the Acknowledgements section:

Acknowledgements

The authors would like to acknowledge the in-kind support provided by Henry Wiersma.

> Any information that is used for the research but that comes from the work of someone else should be referenced. Conference papers typically number references in the order they appear in the paper, while journal papers typically list references in alphabetical order. Check the formatting standards of the conference or journal to which you plan to submit.

References

[1] RILEM Test Method II.4, Measurement of Water Absorption under Low Process, Bagneux, France.

[2] ASTM C 140, Standard Test Method for Sampling and Testing Concrete Masonry Units and Related Units. West Conshohocken, PA, 2012.

[3] C.C. Hamilton, To determine the mineralogical and chemical composition of the Clay Fraction of a hand-grab sample submitted by Mr. H. Wiersma. LR Report: CA01812-FEB05, SGS Lakefield Research Limited, Lakefield, ON, 2005.

[4] Information on http://www.graymont.com/sites/default/files/pdf/bondcrete_masons_brochure_5-01.pdf. Accessed July 4, 2015.

[5] Information on http://www.pacificenzymes.com/. Accessed July 4, 2015.

[6] Information onhttp://www.fyfeco.com/Products/□/media/Files/Fyfe/2013-Products/ Tyfo%20SEH-51A%20Comp.ashx. Accessed July 4, 2015.

[7] Information on http://www.prosoco.com/products/sure-klean-weather-seal-siloxane-wb-concentrate. Accessed July 4, 2015.

[8] ASTM C 1314, Standard Test Method for Compressive Strength of Masonry Prisms. West Conshohocken, PA, 2012.

[9] ASTM D 559, Standard Test Methods for Wetting and Drying Compacted Soil-Cement Mixtures. West Conshohocken, PA, 1996.

[10] K. Mak, C. MacDougall and A. Fam: submitted to Journal of Materials in Civil Engineering (2015)

[11] K. Heathcote, E. Jankulovski, Aspect ratio correction factors for soilcrete blocks, Australian Civil Engineering Transactions. CE34(4) (1992) 309-312.

[12] K. Mak, C. MacDougall and A. Fam: submitted to Journal of Sustainable Building Technology & Urban Development (2015)

CHAPTER 1.12

GUIDE TO ORAL-VISUAL COMMUNICATION

Written reports vs. oral presentations

In any lab or design course that you take, the experimental results you obtained must be communicated in some format. Often, instructors require not only a written report, but also an oral presentation that demonstrates and discusses the background, analysis methods, key results, and conclusions from your experiment. Oral presentations are a key part of any profession you might pursue, so learning these skills is just as important as learning how to write a report.

At first, you might be tempted simply to copy the text and graphics from your report and paste them into a PowerPoint presentation. Yet a few key differences between reports and oral presentations dictate a different approach:

- *Brevity*
  Presentations are commonly limited to five or six minutes; consequently, presenters must select only the most valuable information to include in their talks.

- *Display*
  Text and graphics must be displayed on a screen whose visual requirements are different from those of a text page. Therefore, presenters should develop special copies of graphs, drawings, and tables in order to use screen space effectively. Issues such as color contrast and font size are also key considerations for the presenter.

- *Balance of text and graphics*
  In a presentation, information can be delivered in three ways: text on a slide; visuals on a slide; and spoken words. You will typically use a combination of these methods when presenting. To convey your points to the audience quickly and effectively, you should take a visual approach and limit the amount of text on your slides.

- *Delivery*
  In a written report, the information is delivered solely by words and images on the page. The reader can start, stop, re-read, and review the information at will. Yet in an oral presentation, you are the one who is delivering the information. The slides are only there to support your spoken words. In addition, your nonverbal communication can often make a larger impression than what you actually say. Therefore, it is vital to develop presentation skills that will help the audience follow your train of thought and grasp the key concepts in your talk.

The first part of this discussion will focus on the issues of display and balancing text with graphics. The second part will give some tips on presentation style. A sample presentation follows the discussion.

The slide sequence of an oral report

Title slide
The first slide of an oral presentation does the same job as the title page of a written report. It displays the name of the report, the name of the presenter(s), that person's affiliation, the date, and any other required information. Slide 1 of the sample presentation is a rough characterization of a title slide.

Outline
The second slide of an oral presentation is often a text slide whose job resembles that of the abstract in a written report. It briefly characterizes the project and the results that will be presented. Generic outlines are not helpful; instead, you should tailor your outline to your topic. Slide 2 of the sample presentation below shows an outline slide that is properly tailored to the topic. Note that very short presentations may not require an outline slide, as this information can be delivered verbally instead.

Motivation and Objectives
The next slide of an oral presentation is used to provide some motivation for the study being described and to convey the experimental objectives. A relevant visual or two can be helpful in illustrating the importance of this topic in a particular industry. Complete sentences are to be avoided; instead, use key terms and phrases to get your point across. Examples of a motivation and an objectives slide are Slides 3 and 4 in the sample presentation.

Apparatus and Procedures
Although it is important to show the experimental setup used, the details should be kept to a minimum relative to a written report. One or two slides may be used to characterize the tools and methods involved in data collection. Slide 5 of the sample presentation describes the experimental apparatus and procedure.

Data Analysis Methods
In a technical talk, your audience will want to know the methods used to analyze your results. In this section, you should briefly explain the key concepts and equations from your data analysis. This section can be difficult to follow, even for a technical audience, so be sure to guide your audience carefully through the logic of the analytical methods. Relevant visuals can be helpful in illustrating highly complex concepts. Slide 6 in the sample presentation shows one way to explain the data analysis method. They also show how to display equations on your slides.

Results and Discussion

As in written reports, results in oral presentations are commonly presented visually. However, large tables should be avoided in oral presentations: the viewer usually does not have enough time to process information in a table while also listening to the speaker. Figures are easier to understand than tables, as long as the figures are well designed. Key aspects of figure design are discussed on the following pages. Slides 7 and 8 of the sample presentation stand in for the many graphical slides that might be used to package and display experimental data.

Conclusions and Recommendations

Similar to the report, the conclusions should follow naturally from the discussion. The conclusions are almost always presented as a bulleted list. Recommendations may be presented as a separate slide or on the same slide. The best recommendations are those that help to improve accuracy by addressing sources of error in the lab, or you may suggest logical next steps in experimentation.

Professional appearance of slides

The sample presentation at the end of this chapter shows how to lay out slides to convey information effectively. Slide 9 offers general tips for legible and useful presentation of words and sketches on slides.

Presenting yourself professionally

Slides 10-11 provide simple suggestions that student speakers should review before delivering a presentation. These suggestions involve both quality-assurance review of slides and a summary of effective presentation techniques.

Production and editing

Presentations are made using projectors. Some of these are good, and others are poor. The lighting in the room can also result in a poorly lit presentation. The following suggestions for slide production are designed to make sure you have a good result even if the projection is poor.

Font size and selection

Font sizes should be kept between 18 point and 40 point. Fonts smaller than 18 point will not reliably display, and fonts larger than 40 point will waste too much space on the slide. A non-serifed font such as Arial, Calibri, or Helvetica is the preferred font type for most presentations, as it projects better than do serifed fonts such as Times New Roman.

Text density

Words should be kept to a minimum, even on slides that describe project goals or conclusions. Complete sentences should be avoided in favor of key words and phrases. For example, in an Objectives slide, one might reasonably use short phrases such as these:

- Measure concentration of NaOH
- Determine relationship between pressure and NaOH concentration
- Develop an empirical model to describe this relationship

Professional quality graphics

Projected graphics need to have a clean and sharp appearance. While errors can sometimes be overlooked in small print graphics, these errors will be expanded to huge proportions by a projection system. It is thus best to prepare drawings, tables, graphs, and equations using computer drawing and graphing tools.

Font emphasis

It is usually unnecessary to give your words extra emphasis through use of visual markers such as boldface, italics, and the like. Audiences may find such visual markers to be distracting. Presenters should avoid heavy use of such visual markers; instead, speakers should create emphasis using vocal tools such as volume, tone, and pauses. You may also use a laser pointer to emphasize items on a chart, but do not overuse the laser pointer, as it can be a distraction.

Colors and backgrounds

In some cases, color can help distinguish between different series of data in a figure or can simply add visual interest to a slide. However, color should be used sparingly to avoid garish slides. Poorly chosen colors can accidentally obscure important information, and even the best color decisions are quickly lost if a black and white photocopy must be made. Additionally, about 10% of the population is color blind, so relying purely on color coding to distinguish symbols on a graph is not wise.

To avoid these problems, you should follow these guidelines for slide design:

- Background colors should be uniform, light, and dull. White is an excellent background color for informational slides.
- Foreground colors, such as those in drawings and graphs, should be dark so that they contrast well with the background. Black is an excellent color for the lines in technical drawings and graphs.
- Distinctions are best shown by variations in line weights, line breaks, and the like. However, if color is needed, be sure to choose contrasting colors. Even if a yellow line on a white background looks great on your computer monitor, the resolution will be much lower when projected onto a screen, and the yellow will not be visible.

In short, you should exhaust your black-and-white visual resources before you begin to integrate colors into your graphics.

Horizontal space

A projection screen provides a great deal of horizontal space, and you should strive to fill that space with information. Drawings and graphs should be sized to fill the screen from left to right; when a graphic does fill the horizontal space on the screen, several blank lines of vertical space will usually remain above and below the graphic. This space is best reserved for slide titles, placed above the graphic, and for captions and legends, to be placed below the graphic.

Fonts on figures and tables

Most figures and tables are prepared using one piece of software, such as Matlab, Kaleidagraph, or Excel, before they are placed in a second piece of software, such as Powerpoint. Unfortunately, the font sizes originally selected for the axis labels and legends may not survive the transition, leaving these labels illegible when projected during a talk. To complicate this problem, there is no easily available table of font translations which will help the student to select, say, a Matlab font size that will scale to 20-point during projection. Very simply, the student must conduct a few calibration tests with each piece of graphing software and each piece of presentation software, setting legends, labels, and titles to 20-point, and then adjusting by a few points until the graph or table projects legibly. While this process is labor-intensive, it can be dispatched quickly and relatively painlessly, and it must be repeated each time a new piece of software is integrated into the author's toolkit.

Figures and tables

Data can typically be presented as either a figure or a table. However, figures and tables do different jobs for the speaker and for the audience; thus, it is usually not necessary to present all data in both forms, and authors must decide which form of presentation is appropriate for a given data set or for a given project. In making such a decision, authors must consider the strengths and weaknesses of each form of data display. Tables permit great specificity, as they offer a simple and detailed view of data and of the operations that are performed on that data. However, some people find it difficult to spot trends in tabular displays of data, and most people find it difficult to track and compare trends in multiple data sets when they are presented. Figures support comparisons between sets of data, benchmarks, and the like, although in doing so, they often sacrifice specificity, as it is generally impossible to determine the X and Y coordinates of a given point with visual examination of the plot.

In professional notebooks, figures and tables are likely to be kept together. For quick display in presentations, it is best to avoid tables in favor of figures. It is also wise, however, to place important tables on slides and keep those tables handy during the presentation; because these tables provide comprehensive views of data, they can provide necessary details during the question-and-answer session.

Photographs

When an apparatus must be illustrated, students should provide drawings rather than photographs. When a photograph is used in a presentation or report, that photograph should be aligned with a drawing that schematically represents the important elements of the device or apparatus of interest. In such a circumstance, the photograph should be used in support of the drawing, as professional drawings are more powerful and valuable tools for engineers. Additionally, photographs raise two simple problems that most technical professionals cannot solve alone: it is difficult to take a technical photograph that isolates your device from surrounding clutter, and it is difficult to manage lighting so that your device displays fully in a photograph.

Speaking for figures

After the slides have been developed, it is still the job of the speaker to describe those slides to the audience and to explain what point is made by each figure or drawing. While work on the presentation begins with development of good slides, the presentation itself still depends on the words that the speaker says. No speaker should assume that a slide's point is obvious to the audience; you must always go on record by stating the points that your slides illustrate. Those slides do not speak for you; rather, you must speak for those slides.

Slide 1

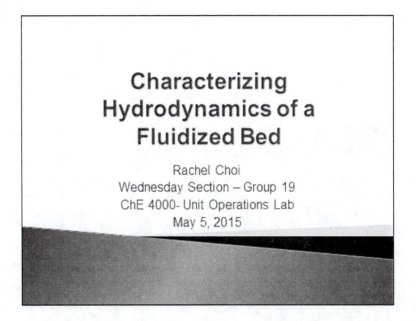

Slide 1 shows an example of a title slide. Text lines are typically centered on title slides. Type size should be 20-point or greater.

As on the cover sheet for a written report, the descriptive title of the report is placed toward the top of the slide, followed by the presenter's name and the institution affiliation or the course title, and the presentation date. Students may be asked by their instructor to include extra information, such as the section number and the instructor's name.

This slide raises the question of orientation. Instructors and professional audience members do not mind whether students prepare slides in portrait or landscape orientation so long as the slides are consistent. Changing the variation for one slide may be okay if necessary to present a particular chart, but generally, landscape slides are preferred for technical presentations. **Most presentation programs offer landscape-oriented slides as a default.**

Slide 2

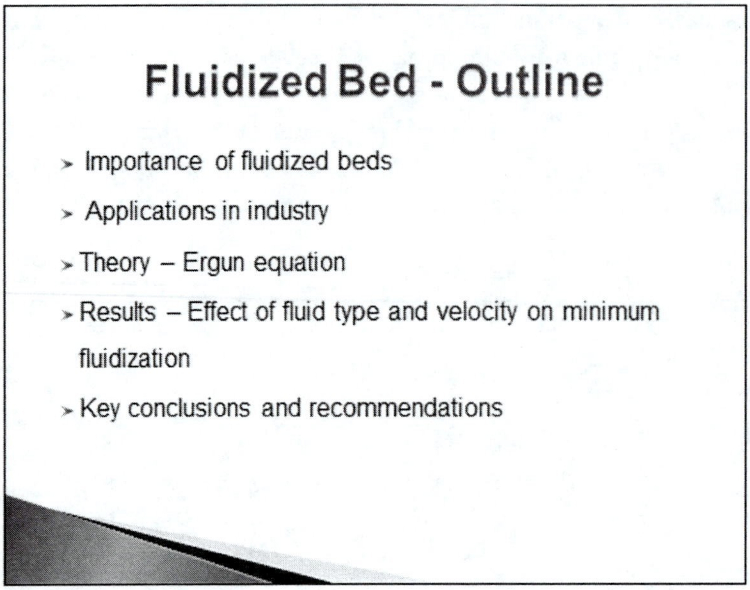

Fluidized Bed - Outline

- Importance of fluidized beds
- Applications in industry
- Theory – Ergun equation
- Results – Effect of fluid type and velocity on minimum fluidization
- Key conclusions and recommendations

The second slide in the presentation should present an outline of the information in the talk. This outline slide, as represented by Slide 2, lists the major format headings and indicates what information will be provided in each section of the talk.

Generic outlines should be avoided. The audience already knows that you will present an Introduction, Theory, Procedure, Results, and Conclusions, so simply listing those as bullet points is not useful. Instead, you should tailor your outline to the topic at hand so that it foreshadows the main points to be covered. For example, in the slide shown here, each bullet concisely indicates the main point of each section of the talk.

Students should present this information in roughly 15 to 30 seconds.

Slide 3

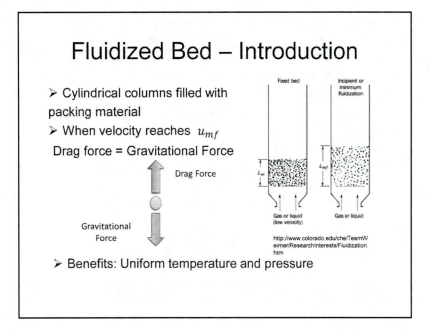

Slide 3 represents an introduction slide. The introduction slide may provide the motivation for the experiment and describe the importance of what is being studied. Relevant visuals may help illustrate the points discussed on this slide.

Students should present this information in roughly 30 seconds.

Slide 4

Fluidized Bed - Objectives

> Determine the hydrodynamics of a water- and air-fluidized bed

> Calculate the minimum fluidization velocity from experimental data

> Compare experimental pressure drop results to Ergun equation predictions

Slide 4 represents a statement of experimental objectives. The project objectives should be stated concisely and specifically. In making this statement, students should avoid showing complete sentences on their slides. Instead, students should display key verb-noun phrases. The verbs will characterize the things that experimental professionals do, such as measure, characterize, develop, compare, design, determine, analyze, test, or verify. The nouns specifically name the technical concepts, tools, or ideas that are analyzed, verified, used, tested, determined, or developed.

Students need to present this information in about 30 seconds.

Slide 5

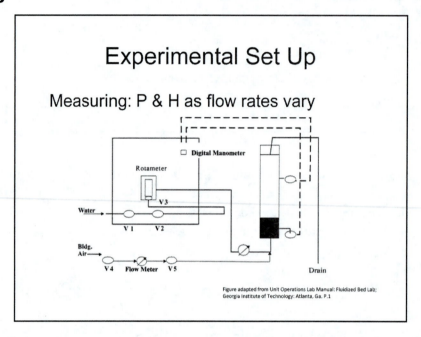

Slide 5 shows a simple sketch of a fluidized bed system, along with a brief description of what was measured (here, pressure (P) and bed height (H).) A sketch of this quality and detail can be used to convey the general layout of an experimental apparatus. Simple sketches like this are helpful because they strip away unnecessary detail in order to convey large structures. If finer views are necessary, they can be presented on the next slide. In general, descriptions of experimental systems should isolate just the relevant structures, such as structures that move, are adjusted, or respond to inputs.

Note that photographs should be used cautiously and should only be used as a supplement to a schematic or diagram of the apparatus. Just from looking at a photograph, the audience usually will not be able to understand exactly how the system is set up. Moreover, the lighting in the laboratory may be poor, and thus the photograph may be hard to see when projected. A slide such as this should be described in 30 to 45 seconds.

Slide 6

Critical equations needed to characterize fluidization

➤ Ergun Equation:

$$\frac{\Delta P}{L} = 1.75\, \frac{(1-\varepsilon_m)}{\varepsilon_m^3}\, \frac{\rho_g u_0}{\phi_s \overline{d_p}} + 150\, \frac{(1-\varepsilon_m)^2 \mu}{\varepsilon_m^3}\, \frac{u_0}{(\phi_s \overline{d_p})^2}$$

➤ Minimum Fluidization Velocity:

$$\left(\frac{d_p^3\, \rho_g(\rho_s - \rho_g)g}{\mu^2}\right) = \frac{1.75}{\phi_s \epsilon_{mf}}\left(\frac{\overline{d}_p\, u_{mf}\rho_g}{\mu}\right)^2 + \frac{150\,(1-\epsilon_{mf})}{\phi_s^2 \epsilon_{mf}^3}\left(\frac{\overline{d}_p u_{mf}\rho_g}{\mu}\right)$$

Equations from Unit Operations Lab Manual: Fluidized Bed Lab; Georgia Institute of Technology; Atlanta, Ga, 2013.

Slide 6 represents the reasonable and appropriate display of an equation. Two simple points are of interest here. First, equations should not be crowded together on slides. Generally, only a few equations are important to a project, and only these important equations need to be shown in a presentation. Second, the equations here are large and easy to read. The size menu in Equation Editor can be used to adjust the font size of the terms, superscripts, and subscripts. The terms in the equations should be defined orally, or if there are only a few terms, they can be defined on the slide.

Students should describe a slide such as this in roughly 30 - 45 seconds.

Slide 7

Theory underestimated u_{mf} – why?

Sources of error:
- Manometer error
- Cohesive/Frictional forces
- Ergun equation neglects wall effects

Fluid	Exp. u_{mf}	Theor u_{mf}	Error (%)
Water	0.063 m/s	0.022 m/s	186
Air	1.14 m/s	1.09 m/s	5

Slide 7 shows how a table and relevant commentary might be placed on a slide. Of interest here are the table's title, size, simplicity, format, and its management of numbers.

Title

Take advantage of the title of a slide as a way to hint at the significance of these results. In essence, a good slide title contains the key question or key conclusion of that slide. A generic title for this slide might be "Results"—which is less informative. Instead, the title actually foreshadows the importance of the data that are shown.

Size

This table has only 12 cells, including column heads. Because there are few cells, the type size can be kept at 20 point or greater, the minimum for visibility on a presentation slide.

Simplicity

Complicated tables should be avoided in oral presentations because the audience does not have time to process them. Instead, data should be carefully selected and arranged to convey a particular point. In presentations, good tables commonly summarize the information that is captured in much larger and more comprehensive experimental spreadsheets.

The table shown in Slide 7 is not comprehensive and does not strive to be so. Instead, it aims to show the discrepancy between an experimental and a theoretical value. The commentary, which may be placed either above or below the table, serves two purposes: it provides more explanation of the table, and it helps the student remember what to say when delivering the slide. Note that the commentary should be limited to key words and phrases; complete sentences should be avoided.

Format

The alignment of the cells should be consistent. The numbers themselves are separated by thin, dark border lines around the cells. Each column has a complete, brief header label, showing units when they are available.

Numbers

The columns of numbers are carefully aligned here. The decimal points are vertically aligned in the *middle* column, and the values are center-aligned in the *Enthalpy (mf)* column. Further, the display of significant digits is both consistent and credible.

A slide such as this should be described in about 1 minute.

Slide 8

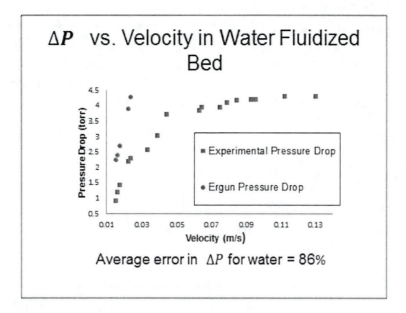

Slide 8 shows a graph as it might be displayed in a presentation. Note that hand-drawn figures should not be used in an oral presentation. Instead, use Matlab, Excel, Kaleidagraph, or another graphing tool.

When emphasizing trends, graphs are more useful than tables. Of interest here are the graph's size, its use of lines and markers, the appearance of the background, and the display of axis labels and legend.

Size

The graph in Slide 8 is sized to fill the slide from side to side. This large display makes it easier for audiences to read the labels and legend, which are provided by the graphing software (e.g., Excel) rather than PowerPoint.

Lines and markers

Each mark on a graph is meaningful, so all the marks on the graph must be visibly distinguished by manipulation of line weights, breaks, and marker shapes. Information on this graph is represented in two distinct forms. Small squares represent the experimentally obtained data. The circles represent values obtained from a correlation, calculated by the students using experimental data and pertinent equations. Additionally, color coding is used to make that distinction more obvious. Note that when plotting values calculated from a theoretical model, students may wish to use a solid line to represent the model, whereas experimental data are best represented as points. You generally should not "connect the dots" of experimental data.

Background and grids

Data are best observed when they are displayed with dark lines and markers arranged on a light background. The white background of this graph provides excellent visibility for the markers This graph also demonstrates that grid lines are largely unnecessary, as the key trends are understandable without any grid lines. Graphs are best used as tools to display trends in data and to compare experimental data with predictions from theory or correlations.

Axes and legends

In this graph, as in all good graphs, both axes are fully labeled, and units are shown in each label. A legend is also clearly visible, providing a key to the different markers and line weights that are used on the plot. Note that the font size on the axes should also be no less than 20 point.

A slide such as this should be described in roughly 1 minute.

Slide 9

Slide Design Concepts

- ➢ Avoid clutter
- ➢ Use relevant visuals
- ➢ Have one main point per slide
- ➢ Key words and phrases only
- ➢ Avoid pointless animation
- ➢ Use contrasting colors
- ➢ Keep font style consistent and at least 20 pt.

Slide 9 provides simple, summary tips on effective slide design. Clutter in technical slides is best avoided by designing figures and charts according to the principles described above. The use of relevant visuals can also help the speaker convey information using few words.

Text slides can be kept free from clutter by sticking to key words and phrases only. Complete sentences are to be avoided on all slides; after all, the speaker needs to talk during a presentation and should never simply read the slides to the audience. A speaker needs to be prepared to add value to the visible information on the slides.

PowerPoint offers many animation options, but these should be used with caution. Most animations will simply distract your audience, so only use it when necessary. For instance, if your goal is to show how something was built, then you may choose to use animation to demonstrate the steps required for its construction. But when switching from slide to slide, you should generally avoid animations.

As noted earlier, colors can be helpful to distinguish points in a graph or to add visual interest to a slide, but be sure to choose contrasting colors. Finally, you should choose a font that projects well, such as Arial or Calibri, and use it consistently throughout.

Slide 10

Preparation

➢ Rehearse and Time; Respect Time Limit

   ➢ Allow for questions, interruptions

➢ Check Spelling and Drawings

➢ Check for Internal Numerical Accuracy

➢ "Sanity Check" Your Results

➢ Preview Presentation Room

➢ Verify Equipment / Supplies

Slide 10 addresses preparation, and these steps require that speakers get help from their colleagues or team members. Presentations should be rehearsed aloud, with one student tracking time while another student speaks. Students should have their friends review their slides and illustrations in search of obvious errors and typos. You may also ask team members to review your slides; ask them to look for numerical inaccuracies or flaws in preparing equations, table heads, graph labels, and so forth. When performing a sanity check on your presentation, always ask yourself, "Do these results make sense?" And, finally, speaker(s) should try to examine the room where a presentation will be delivered, to determine whether the projection equipment works, how it works, and whether the slides that have been prepared will really work on the equipment in that room. Be especially attuned to problems that can occur when slides created on a Macintosh computer are displayed on a PC. Arriving 10-15 minutes early will give you some time to address these issues and will help you to relax without feeling rushed.

Slide 11

> ## Effective Speaking
>
> ➤ Use professional, technical language - Know your audience
>
> ➤ Make eye contact with your audience
>
> ➤ Vary your pace and tone - Don't memorize your talk
>
> ➤ Use spoken transitions between slides
>
> ➤ Speak loudly enough to be heard
>
> ➤ Keep your hands at your side, and gesture occasionally
>
> ➤ Avoid distractions and filler words
>
> ➤ Limit use of laser pointer

Slide 11 provides some fundamental tips for presenters. Very simply, presentations should be taken seriously, and speakers are expected to deliver talks that are formal, specific, and complete. The level of formality may vary somewhat, depending on the audience and the context; for a class presentation, however, the presentation should be formal.

In addition to the words on your slides, your eyes, voice, posture, and gestures communicate volumes to the audience. Making eye contact with the audience rather than staring at the screen is the best way to engage your listeners. Try to look at one person for a complete thought, and then look at someone else in the room for your next sentence, rather than trying to "scan" the entire audience with your eyes!

Your voice is also a powerful tool in your presentation. While you must be professional and technical, you should still try to adopt a conversational, natural tone. Do not try to memorize your talk. Memorization can make your speech seem unnatural; moreover, if you try to memorize and then suddenly lose your place, you can get quite flustered.

The pace of your talk should be relatively slow, but may also vary--you can try slowing down a bit on important points and speaking more quickly when delivering less important details. Insert spoken transitions between your slides so that the audience can follow the logic of your organization. Your voice should be loud enough so that you can be heard at the back of the room. When practicing, ask your friends or classmates to listen for distractors and filler words such as "um," "like," "uh," and so on. Even polished speakers use filler words from time to time, so you probably will not be able to eliminate them. But you can try to minimize them by pausing briefly rather than saying "um." Get used to inserting brief pauses in your talk; as long as the pauses are not uncomfortably long, your audience won't really notice them, and they will give you a chance to gather your thoughts.

Finally, the laser pointer should be used judiciously. It is best for pointing out particular values in tables, or for emphasizing a specific line or trend in a graph. When overused, however, the laser pointer just becomes another distraction. Avoid using the laser to underline all of your bullet points, and certainly do not swirl the laser aimlessly on the slide. Again, watch for these tendencies and try to minimize them during your presentation.

Posters

In addition to written reports and oral presentations, posters are another common medium to report experimental results and conclusions. You may be asked to create a poster for a course or for a poster session at a conference if you are working in a research laboratory. Templates for poster design are widely available; these provide formatting assistance to help beginners quickly prepare acceptable displays. Professional conferences commonly provide templates--usually pre-formatted Powerpoint files—for this purpose, and some university departments maintain an archive of useful templates. Faculty members who are active researchers typically maintain a set of poster templates for their students' use; these are usually pre-formatted with university and department colors, emblems, and the like. If you are unable to obtain a poster template from a convenient source, you will be able to locate numerous free templates by simply using a search engine.

Once you have obtained a format template, you will need an example poster to help you start designing your own display. To fill this need, we present an example of a poster that might be created for an experiment performed in a lab course.

Ten Tips for effective poster creation and presentation
1. In creating a poster, your goal is to display your key points in a way that could easily be understood by a viewer without having the authors there to explain it. The audience for a poster is typically not specialists in the field; however, the audience will usually have a background in science or engineering.
2. PowerPoint or Adobe In Design may be used to create your poster. As with slides, the visuals on your poster should not be hand drawn.
3. Color may be used to add visual interest or to color code figures or equations, but be sure to use contrasting colors and do not use more than 2-3 colors on a single poster.
4. Each section of the poster should be short and may contain a combination of visuals and text. The sections should not be too densely filled with text—imagine a viewer looking at your poster for 30 seconds to a minute. Therefore, large blocks of text should be avoided. The main focus should be on conveying your main points and bottom line in a brief, clear, and visually engaging manner.
5. Relevant visuals should be included—but be sure to cite any image that you take from the Web, textbook, or another source.
6. Put the most important things in a spot where they will be seen:
    a. Most people will look at the upper left corner first.

b. Most people will also look at the middle of your poster. They will glance at the tops of your columns/sections, but may not read all the way down.

c. Keep in mind that in "real life," people will be standing when looking at your poster.

7. When appropriate, you may use key words, phrases, and clauses rather than complete sentences. Bulleted/numbered lists are also useful.

Example: higher pressure $\rightarrow$ higher flux: greater driving force vs. concentration gradient. Still, the text that you do use must be free of errors and typos.

8. Remember that the text should be readable from about 2 or 3 feet away, so 20 point font or larger is recommended.

9. Most of your audience will be interested in your figures and charts, so be sure they can read the font on the axes, labels, and legends.

10. If participating in a poster session, you should develop an "elevator speech" that you can give repeatedly as viewers approach your poster. The elevator speech should convey the main point of your experiment in no more than 30-45 seconds.

Visuals are helpful, not distracting. The font is >20pt; only a few colors are used; and color contrast is preserved by using dark colors on a white background.

The title of your poster, your name, and your institutional affiliation (or course) number/title) should appear at the top of the poster.

Each section of results is labeled with a descriptive heading (e.g., Murphree Tray Efficiency) so that the viewer can follow along.

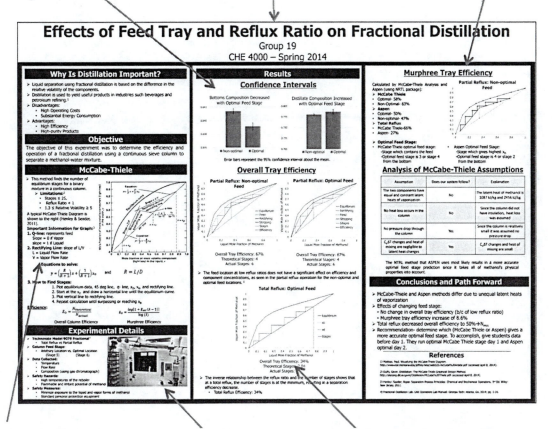

Use key terms, phrases, drawings, and symbols instead of long sentences to explain theoretical concepts.

Photographs of the apparatus can help put your results into a tangible context.

Key results are displayed visually using bar graphs and figures. Large, complicated tables are avoided. Results are displayed in the middle of the poster so that the viewer will notice them easily.

Part Two:

Standards for Undergraduate

Engineering Reports

CHAPTER 2.1

INTRODUCTION

Producing an engineering report is a design and production project just like any other engineering project. The business of composing a report may not be as interesting as the research or development that it documents; however, the report can be equally important because innovation requires change, and change will not happen unless inventions and discoveries are communicated. This second part of this text on engineering reports discusses the job of producing a report. Here, some ideas will be presented about applying the tools used in engineering projects—planning, design, and implementation—to the preparation of a technical report.

Plenty of successful projects in every field are completed without any apparent planning or organization. The participants just rely on luck and hard work. A lot of time and effort will probably be wasted, but if the project is successful, the waste will probably be forgotten. The same can be true of the minor task of producing the report that documents the project, but here the wasted time and effort and the possibly less successful outcome seems particularly unfortunate for at least a couple of reasons. First, the engineer should be a professional with respect to design, research, and development, but he or she is probably an amateur when it comes to report preparation. The seasoned professional can rely on experience and ingenuity, but the amateur should employ planning and standard methods. Second, the engineer should save his or her energy and resources for the most valuable work he or she can contribute, which is engineering, not writing. So, consider an engineering report to be an engineering project that can be done much more efficiently and effectively by using the engineering method.

Every engineering project, like almost everything—except maybe a Mobius strip—can be seen to have a beginning, a middle, and an end. Most projects are organized at least implicitly around the three corresponding stages of conceptual design, preliminary design, and detailed design. In the conceptual design, the goal is defined and the scope and general features of the design solution are envisioned. In the preliminary design, the major components are identified and integrated into a cohesive system. In the final detailed stage of design, the construction documents and detailed specs are generated. The same organization works well for report writing. This part of your text is organized around these three stages.

The first stage of conceptual design is definition and invention. The analogous stage in report writing is defining the type and scope of the report that is required or suitable, the appropriate overall design or format, and the general content or outline. The next chapter addresses these issues.

The next stage, or preliminary design, is when the project engineer exults and excels. This is the stage of selecting the components and putting them into functional subsystems, then connecting the subsystems into effective systems, and ultimately integrating the subsystems into an optimal whole. The analogous stage in report writing is developing the first draft. This is the time to flesh out the outline with well structured paragraphs of effective sentences that present the technical content. Most of the technical content will actually be presented in exhibits such as graphs and tables. Sections 2.3-2.16 in this part address the general design of these components and subsystems.

The final stage of design is putting the finishing details in order and assuring good quality control in writing and production. The final chapters of Part II of this text are devoted to the details of these important issues.

CHAPTER 2.2

WRITING CONVENTIONS IN TECHNICAL REPORTS

You write reports so that people can understand your work by reading the report. You can best do this if you understand how readers search for information in reports. In this section we show how they do this. Here, we will describe the conventions for the preparation and page layout of technical and scientific reports. Following this we will explain, with examples, the mechanics of clear sentences and paragraphs.

In our classes, we use these writing conventions because they are widely accepted and because they enable us to be fair and consistent when grading your reports. However, no document convention is absolutely universal; in your career, you will encounter supervisors or clients whose preferences differ from our guidelines, and you will wisely accept these new requirements. We present our guidelines as default settings for your professional reports. Like any set of defaults, these are designed to offer reasonable performance across a broad range of situations and to enable you to adapt to new requirements with a minimum of difficulty. So long as you understand how to use your editing software, adherence to these document conventions will make it easy for you to adapt to any new requirements that your clients or supervisors may impose on you.

Page layout settings.
- Use 11 or 12 point Times New Roman font, 1.5 line spacing and full justification
- Set margins to 1.0 inches on all sides of each page
- Number the pages of your reports
- Set report titles and section headings in bold
  - o Titles should appear alone and centered on a line
  - o Section heads should appear alone and left-aligned on a line
  - o Side heads should appear as the first words of a paragraph
- Minimize blank space above and below headings and displays
- Indent paragraphs and minimize blank space above and below paragraphs

References

In your reports, you should prepare reference entries to document any values that you did not measure and any displays or explanations that may have been published elsewhere. References should be compiled in a headed References section that immediately follows the headed Conclusions section of your report. The References section may be placed on the same sheet as the last lines of the Conclusions section, with no additional space introduced between the running text and the heading. Chapter 2.6 offers a more complete explanation of reference format and the ethics associated with source documentation.

Spelling out numbers

Here are some guidelines to follow when determining when to spell out numbers in your reports and when to write them as numerals:

1. Numbers less than 10 should be spelled out; numbers 10 and greater should be written as numerals (e.g., The experiment used 30 flasks. We then poured the contents of the flasks into two tanks.)

2. However, any number associated with a unit should be written as a numeral, regardless of how large the number is (e.g., The pump was run at 5 RPM.).

3. Put a space between the numeral and the unit, except with ° and %, where there is no space between the numeral and the unit.

4. Use all numerals in a series or range containing any numbers 10 or greater (e.g., Either 5, 8, or 12 experiments were run.).

5. When a sentence starts with a specific quantity, spell out the number as well as the unit of measure (e.g., Fifteen milliliters of supernatant was then added to the vessel.) If possible, however, rework the sentence to avoid starting with a number (e.g., Then, 15 ml of supernatant was added to the vessel.)

Appendices and attachments

In most class reports, displays are integrated into the body of the report and described in text adjacent to the display. However, some larger report elements may be placed at the back of the report as Appendices, attachments or annexes. Such attached displays may include, for example, spreadsheets, mathematical derivations or intricate calculations. The attachments in an appendix usually represent work that is germane to the project results but whose complete presentation may digress from the results presentation. In general, lengthy displays are attached to reports in order to fully disclose all of the work that was done on the project; these are relegated to an Appendix when they are judged not to be immediately critical to the points made in the body of the report.

A word about proofreading

When you submit a report to a client or supervisor, those readers assume that the report demonstrates your standards as a professional. This assumption extends to simple document assembly issues such as proofreading and correct assembly of the document. Documents that contain silly proofreading errors are assumed to contain equally silly errors in, for example, data collection or calculation.

You should proofread your documents and check all of your calculations and measurements for accuracy. But you should not do this alone—you need to get your colleagues and teammates to review your work as well. You should always assume that your documents contain errors that you can no longer see; your teammates should help you locate these errors, just as you should help your teammates to do so in reports that they prepare.

Page layout with displays

Because the bulk of your project results will be presented as tables or plots, it is useful now to outline how to properly insert these visual displays into a formatted page of your text. When we speak of displays here, we are primarily talking about graphs, tables, sketches and, perhaps, sample calculations, as these are used frequently in undergraduate project courses. Other types of displays will be discussed later in the book. In the model reports here, displays are integrated into the body of the report; that is, displays are placed adjacent to their text descriptions. From time to time, however, instructors will ask you to place your displays at the back of your reports, as attachments. When this happens, the text citations and descriptions of the displays will not change. Only the location of the displays will change in such circumstances.

In the remainder of this section, we outline the norms for displaying figures, tables and sample calculations.

Figures

The figures in lab reports are typically graphical representations of data, as are Figures 1–4 in the sample "Lab Zero" report of Section 1.3. Many people loosely speak of such displays as *graphs* or *plots*; these terms are not exactly wrong, but they are informal and can lead to confusion. In reports, most displays will be cited and numbered as *figures*, and you should adopt that terminology in your work.

Every figure in your report must have a figure number and a brief descriptive caption. The figure number and the caption are always placed on the same line, below the figure. The figure number is placed to the left of the caption. On rare occasions authors are allowed to save space by using a single caption for a multiview display—a group of closely related figures. You should consult with your instructor before creating multiview displays for your courses, as these can present problems with layout and visibility.

The axes in plots must be labeled, and each label must show the units used on that axis. You should include a legend with each plot to define all the information on the plot. When numerous data sets or models are displayed, the legend distinguishes the different markers and lines. In all figures, axis scales should be adjusted to frame the data tightly and reasonably. Significant blank space should be avoided on plots; if your measured data ranges from 0 to 75, then your axis should range from 0 to 75, not from −50 to +150.

When you introduce a figure, you should always first *cite* the figure in the text just above the point of display on the page. In student reports, figures are commonly inserted just below the paragraph in which they have been cited and described. A figure *citation* formally directs the reader's attention to a particular display, using the figure number as if it were the name of the figure, as here:

> The results are shown in Figure 1.

As in the citation above, the figure number is always capitalized and spelled out fully.

Figures should be numbered in the order that they are introduced in the report; the first figure you cite should be "Figure 1," the second figure should be "Figure 2," and so forth.

When figures are attached at the back of a report, they must still be numbered and arranged in order of citation.

Because data plots present your data, you need to format them in a way that fully presents your information while avoiding ambiguities. To do this, you first need to attend to simple matters such as selecting an appropriate scale, labeling the axes with units, and using fonts that are sufficiently large to enable readers to read them easily. It is best to set the fonts in your displays to be the same size as the surrounding text in your reports. In addition to setting fonts appropriately, it is important to use data markers that display distinctly, and to provide a legend that defines your markers. It is best to select markers that display well in black and white, as instructors and graders will often print your reports on black and white printers, eliminating color distinctions that were introduced by your plotting program.

The plot area of your graphs should be free of lines other than the axis marks. Plotting programs allow you to introduce lines to connect your data points. You do not usually need such lines when you have only one data set to plot. You may use *straight* lines to group (connect) data markers IF you have plotted multiple data sets that must be kept visually distinct. Lines that are curved or *smoothed* represent mathematical calculations such as models or trendlines; it is assumed that all the points on a curved or smoothed line are known. If your plot displays a trendline or model line, you should display the equation for that line as close to the line as you are able. Finally, on some projects, you may be seeking a particular data point, such as an inflection point; you can and should label such points using any drawing or annotation tools that your system provides you.

The example plot below illustrates how to make a display that is clear and easy to read.

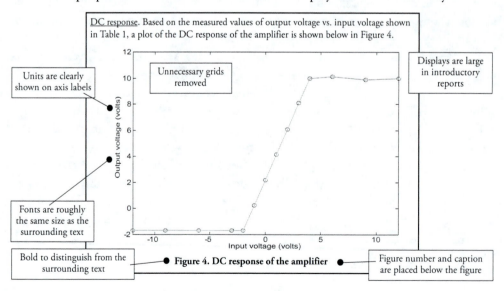

Tables

Tables are to be treated with the same care and precision as figures. The rows and columns in your tables must be clearly defined and labeled, and units must be visible either at the tops of the columns or at the left edges of the rows. Like figures, tables must have numbers. Unlike figures, tables have titles that appear above the table. The table number should be placed to the left of the title. As is the case for figures, tables must be placed and numbered sequentially and in the order of use. Tables should be cited in the report text *before* they are shown on the page, and specific tables must be referred to with a capital letter "T."

When you present tables in the body of a report, those tables should be small; they should have few rows and columns, and they should present results that the reader needs to remember. Large tables, displaying raw measurement data and calculations, are best attached as appendices to your reports. This allows readers to examine your measured values as needed after they have reviewed your results and conclusions.

Table 2 of our sample report is reproduced here. It demonstrates how to number, title, and label tables; please notice that the last sentence before the table is introduced is designed only to cite the table, as shown by the statement that introduces the sample table.

Table 2 shows the measured input and output voltages with the calculated gain.

Table number and title are usually bolded and placed above tables

Entries are aligned on the decimal or other repeated element

Table 2. Frequency response of the amplifier

Frequency	V_{in} (volts)	V_{out} (volts)	K	Gain (db)
10	1.0	5.0	5.0	14.0
10^2	1.0	9.8	9.8	19.8
10^3	1.0	10.1	10.1	20.1
10^4	1.0	10.0	10.0	20.0
5×10^4	1.0	10.1	10.1	20.1
8×10^4	1.0	10.6	10.6	20.5
9×10^4	1.0	10.7	10.7	20.6
10^5	1.0	10.1	10.1	20.1
2×10^5	1.0	6.0	6.0	15.6
10^6	1.0	3.0	3.0	9.5

Presenting Calculations

In order to obtain an analytical result, or derived value, you must transform your measurement data. You do this using mathematics. When you display an analytically derived value, you must document the steps by which you transformed the measured data. Because calculations are routine in the sciences, you are expected to be able to account for your calculations clearly and efficiently. Here we briefly explain how to do this.

In general terms, calculations are presented in the same way as other technical procedures: you describe the goal of the calculation, account for the inputs to the work, describe the methods you used and account for the result that was obtained. In describing a calculation, the goal is the quantity you wish to obtain at the end of the calculation process, the input to the work is the set of measured values that you use for the calculation, the methods include the equations you used with definitions of the variables, and the result is the value obtained from the calculation.

Following these principles, simple calculations can be presented in a four-step process:

1) State the goal of the calculation
2) Cite the equation
3) Display the equation in symbolic form
4) Define the variables and input values

Shown below is a sample calculation pertaining to an Oscilloscope investigation. Please notice that this example devotes three lines to equations, it devotes two text lines to explanations at the end, and it devotes a further three lines to text descriptions of what physical entity is represented by each variable in the equation:

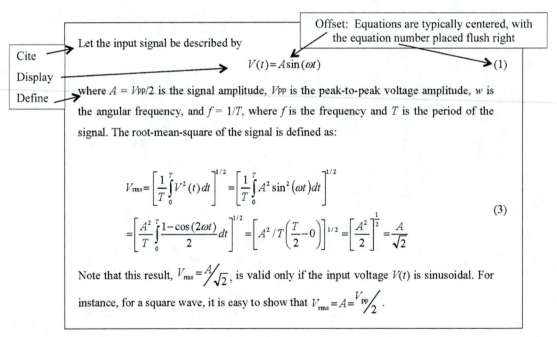

Offset: Equations are typically centered, with the equation number placed flush right

Cite

Display

Define

Let the input signal be described by

$$V(t) = A\sin(\omega t) \tag{1}$$

where $A = V_{PP}/2$ is the signal amplitude, V_{PP} is the peak-to-peak voltage amplitude, w is the angular frequency, and $f = 1/T$, where f is the frequency and T is the period of the signal. The root-mean-square of the signal is defined as:

$$V_{rms} = \left[\frac{1}{T}\int_0^T V^2(t)\,dt\right]^{1/2} = \left[\frac{1}{T}\int_0^T A^2\sin^2(\omega t)\,dt\right]^{1/2}$$
$$= \left[\frac{A^2}{T}\int_0^T \frac{1-\cos(2\omega t)}{2}\,dt\right]^{1/2} = \left[A^2/T\left(\frac{T}{2}-0\right)\right]^{1/2} = \left[\frac{A^2}{2}\right]^{\frac{1}{2}} = \frac{A}{\sqrt{2}} \tag{3}$$

Note that this result, $V_{rms} = A/\sqrt{2}$, is valid only if the input voltage $V(t)$ is sinusoidal. For instance, for a square wave, it is easy to show that $V_{rms} = A = V_{PP}/2$.

Spreadsheets

If a spreadsheet is required in a report, include as an attachment a printout of all important parts of the preliminary spreadsheet. The spreadsheet should be as complete and tested as possible. The spreadsheet should have a unique number and a descriptive title; this information should be centered and prominent on the page, as shown here:

Attachment 1. Experimental Spreadsheet

The heading can then be a mere line in your report text, which will be easy to edit and update.

Prelab exercises often require you to illustrate an important calculation with a block of formulas. If so, append a small block of representative cell formulas on a separate page. Do not print a large block of formulas along with less relevant data, such as numerical data, labels, and so forth. Rather, print a pertinent, representative block implementing the critical calculation or algorithm. Always describe this calculation in an accompanying text paragraph.

CLEAR WRITING IN TECHNICAL REPORTS

Style in paragraphs

In scientific writing, your paragraphs should *define* and *explain* a single topic or concept. Readers expect you to do this in paragraphs that appear to be clear, concrete and orderly. Some instructors will tell you to treat each paragraph as if it were a small technical report, with introduction, results and (sometimes) conclusion sections. This is a good suggestion; clear paragraphs have a well-defined structure with these parts:

1) Topic statement (or topic *definition*).
2) Development or *explanation*, with details.
3) Summary or evaluation point (optional).

In a clear paragraph, the Topic statement should usually be compressed into one sentence. This sentence should define what the reader will learn about in the paragraph, and it may show the logic that will drive the following discussion.

The Development section of a paragraph should introduce details that support the topic or claim. This section should do more than list the items of supporting evidence; in this section, your comments should display the logic that makes the details important. You may also need to include comments on the quality (good or poor) of the data you have collected.

In scientific reports, some important paragraphs provide a concluding Summary or Evaluation. Usually this happens only once in a headed section.

The annotations on the following paragraph show how the paragraph can be subdivided into *topic*, *development* and *summary* sections.

Topic Definition	The goal of this first experimental step was to characterize the behavior of the thermocouples to be used in the rest of the project. Specifically, readings obtained from the three thermocouples that will later be placed in the test samples were compared to the readings from the thermocouple that will be placed in the water bath alone and from the thermocouple that will measure the skin temperature of the container. The three sample thermocouples were arranged in approximately the
Development with details	positions they will take in a spherical sample of the experimental material. They were then placed into the bath and allowed to settle to a constant temperature. The thermocouples detected slightly different readings for the bath temperature. These different readings are shown in Figure 6, which also indicates that the bath temperature thermocouple recorded a temperature of fully one degree Celsius below the reading recorded from the thermocouple that recorded skin temperature. Further, the skin temperature thermocouple showed a temperature Celsius below both the sample midpoint and the sample center thermocouples. Figure 9 displays
Summary/ Evaluation	these results graphically. These results indicate that the thermocouple must be calibrated to agree at a common temperature in order for subsequent sample tests to provide precise results.

Logic and flow of ideas

A paragraph should explain a single topic or concept using sentences that present a consistent flow of ideas. When your sentences all address a single concept, we say that your paragraph is focused on that topic. When your ideas seem to flow smoothly from sentence to sentence, we say that your paragraph is cohesive. Clear paragraphs need to be both *focused* and *cohesive*. In order to assess focus and cohesion in your paragraphs, we examine the <u>sentence subjects</u> and <u>transition statements</u>.

Sentence subjects name the central component or idea in a sentence. When the sentences of a paragraph consistently use a particular topic or concept as the sentence subject, those sentences can be said to *focus on* that topic. *Transition statements* are usually placed at the beginning of a sentence ahead of the subject, and they may be only a few words long. "However," "therefore," and "because," are common single-word transition statements. Such transition statements are used to define explicitly the logic that relates a sentence to the sentence or sentences that precede it.

Transition statements and repeated sentence subjects are tools that writers use to establish flow and focus in paragraphs. It is not necessary to use both tools in every sentence. However, it is wise to use at least one of these in each sentence of your scientific writing. When you do this, your readers will view your reports as short, clear stories about the work of science and about the physical processes you have examined.

Annotations on the example paragraphs below call attention to the repeated words that establish focus and the transition statements that display the author's logic.

Figure 8 shows the (stress) distribution of the [spliced] beam, in this case with an FRP box beam length of 400 mm. The [spliced] beam does not bend with uniform curvature, as the FRP box beam significantly stiffens the beam over the [splice] length. Regions of high (stress) are identified in Figure 8: Location *A* is at the mid-span where the steel web [splice] plates meet the gap between the centre-beam ends; Location *B* is at the end tips of the FRP section. The (stresses) at these l ocations are indicated in Table 2. The results in Table 2 are for 103 mm x 103 mm FRP box sections. Two different FRPs (glass FRP and carbon FRP) were considered, and the thickness *t* of each box section is indicated in the table.

Repeated words: Stress is circled, Splice is boxed

In order to assess the feasibility of the proposed splice detail, a finite element model was developed and investigated. The finite element model was implemented in ABAQUS and is shown in Figure 6. (Although) the centre-beams of MBEJs are typically continous over several spans, it was decided to model only a single-supported span to reduce the size and complexity of the model. (In the future) the effects of continuity will be investigated. the span of the beam is 1 metre, which is typical of the centre-beams for an MBEJ, and pin and roller supports were assumed.

Logical connectors are circled

127

Paragraph Structure
A Quick Reference Sheet

(**Main components are in bold**. Underpinning questions are in roman type.)

Issue (Issues announce topics, raise questions, and make claims.)

    1. What project task or question is addressed in this paragraph?

    2. What claim is made (or What result is presented)?

Discussion (Discussions describe tasks, explain answers, or substantiate claims.)

    1. How is the result obtained?

    2. How is the claim substantiated?

    3. Why was this decision made? (or Why was this conclusion drawn?)

    4. How might this open question be addressed?

Style in sentences

Technical reports should be written with language that is simple, direct, and descriptive. While technical reports address concepts that sometimes seem hard to grasp, these concepts can and should be described using terms that are short, simple, and can be written without internal punctuation. Even the variables that we use in complicated equations have names that describe physical entities or forces, and these names are typically short, simple, and can be written with a small number of keystrokes. Your reports need to be written in such a way that they are easy to read even for an engineer who has not participated in your project. In order to write such reports, you need to pay attention to the way you assemble every sentence.

In technical reports, sentences need to be impersonal and objective. These sentence qualities are directly linked to the subjects, verbs, and modifiers from which all sentences are constructed. The following paragraphs discuss these qualities in detail.

Impersonal structures

Impersonal structures are formed from sentence subjects. In the subject position of each sentence, you answer the question "Who or what is this sentence about?" Members of the research team (that's you) rarely appear in sentences describing their projects. The subjects of your sentences should be the things (or forces) you are studying.

Avoid this: "We found that the pressure varied with changes in temperature."

Do this: "Pressure varied with changes in temperature."

Empirical structures

Empirical structures are formed with verbs. In the verb position of each sentence, you answer the questions "What happened?" or "What did [the subject/agent] do?" The answers commonly involve active verbs—things *deflect*, *break*, *boil*, *cool*, and so forth.

Experiments require observation, so you should expect to use some verbs associated with seeing. Phenomena are *observed*, *seen*, *found*, and *shown*, for example, and values are *calculated* and *determined*. These words represent actions that you take. But because you do not get to appear as subjects in your impersonal sentences, you must sometimes use the so-called passive verb constructions in order to maintain an impersonal stance while recounting your work faithfully. Passive constructions appear occasionally in impersonal texts.

First Person Passive Impersonal, active	**Avoid this:** "We used Equation 4 to determine the Reynolds number." **Do this:** "Equation 4 was used to determine the Reynolds number." **Or do this:** "The Reynolds number was determined using Equation 4."

Objective structures

Objectivity in sentences is maintained with modifiers. The best known modifiers are adjectives and adverbs. With these structures we characterize the things we observe and the changes we observe; modifiers help us to answer the "How" questions that drive science reporting: "How large?" "How many?" "How fast?" In response to these questions, the reader needs to see numbers.

Adjectives can be imprecise Numbers are specific and precise	**Avoid this:** "A large sample was heated to a high temperature for a long time; then it was cooled quickly." **Do this:** "A 70 gram sample was heated to 200 degrees C for 4 hours. It was then cooled to 0 degrees C over the course of 20 minutes."

Managing verb tenses

Many students are confused about the relationship between tense management and passive constructions. We can get past this confusion by remembering one principle: past tense constructions are different from passive sentence constructions. Past tense constructions are used to describe any action taken in the past. When a report describes an experimental investigation, it will be written in past tense because the experiment was performed in the past.

The so-called passive construction is primarily a sentence inversion that places the sentence's *object* in front of the sentence's *subject*. When sentences are inverted in this form, the subject of the sentence is often dropped. Technical reports often use passive constructions in order to maintain an impersonal stance, as here:

Avoid this:
"We used Equation 4 to determine the Reynolds number."
This is first person and it is past tense.

Do this:
"Equation 4 yielded the Reynolds number."
This construction is impersonal, and past tense.

Or do this:
"The Reynolds number was determined using Equation 4."
This is still a past tense statement; because it is also inverted, it is also called passive.

CHAPTER 2.3

GUIDELINES FOR EDITING SPECIFIC FEATURES OF REPORTS

Formal technical reports share many features of page design and writing style. This is because employers, publishers and professional societies use *Style Guides* or *Style Manuals* to define rules for preparation and page layout of documents that are submitted to clients, supervisors or to management. Most professional societies have developed style guides that they require members to use. The American Psychological Association (APA), uses the book-length *Publication Manual of the American Psychological Association*, while the Institute of Electrical and Electronics Engineers (IEEE) uses a modest-sized style manual, *The IEEE Editorial Style Guide*, that is available as a free download from the organization's website. Like most professional societies and journals, IEEE also provides document templates as free downloads from the publication section of its website. These templates encode the layout components of their professional style guide, using files that are compatible with widely-used editing programs such as Microsoft Word and LaTeX. Some professional organizations, such as ASME also provide templates for the Adobe Framemaker editing program.

Document templates are not always available for undergraduate courses. Consequently, we review here the main components of document layout for undergraduate reports, and we suggest default settings that most instructors find to be acceptable. Using these guidelines, students should be able to create and modify document templates that will be useful for a variety of courses in which technical reports are required.

Page design

The most prominent features of page design are margins, justification, and line spacing. These features govern the appearance of your report, and they are easy to modify.

Margins

Uniform margins make your reports look professional, and they make the pages of your reports easy to handle. For classroom reports, you should set your margins to 1.0 inches on the top, bottom, left and right sides of the paper sheet. Alternative settings, such as 1.25 inches on the left and right, are most commonly used when documents are to be printed and then bound. Your class reports are unlikely to be bound, so you can adopt 1.0 inches as a default margin for all sides of your paper sheets.

On occasion, you will find that a large display wraps across a page break, creating a blank space that resembles an unacceptably wide bottom margin. Such blank spaces look unprofessional and are to be avoided. The technical editing program LaTeX manages this problem nicely by automatically adjusting the display's point of insertion in the running text of the document. However, users of other editors must make such adjustments by hand.

Line spacing

Research indicates that single-spaced text is somewhat difficult to read and that readers require more time to read single-spaced text than double-spaced text [1]. This phenomenon probably arises from readers' difficulty in resetting the eyes correctly to the start of the next line when the lines are closely spaced. Although professionally printed material, such as that found in technical journals and newspapers, is usually single spaced, these publications are generally set in multi-column format with narrow columns that facilitate resetting the eyes.

In contrast to journals and newspapers, your classroom reports will be graded, marked and returned to you. Comments are easiest for teachers to insert when student reports are prepared using a single column of text that is a full letter-page wide. For coursework reports, you should set line spacing to 1.5 lines per feed. This setting will facilitate your graders' reading, and it will give the grader room to write comments on your pages.

Justification

Justification describes the point against which a line of printed text is aligned. When text uses *Left-Justification*, all lines of text are set at the left margin, forming a straight line, while the right ends of the lines may present a ragged appearance similar to that created by mechanical typewriters. In professionally printed documents, text lines are usually *Fully Justified*, which means that the left and the right edges of each line terminate at the margin. In fully justified text, such as in this book, the editing program manages character spacing and hyphenation in order to guarantee clean and straight lines on the right and left sides of the text column. For your class reports, you should set the lines to full justification.

Some document components do not fill complete lines. When these are used, students must take extra steps to manage justification. Subsection headings, for example, are usually placed flush with the left margin. Equation numbers are usually placed flush with the right margin.

Paragraph format

Readers do not like to read long paragraphs. Further, when readers cannot tell where one paragraph ends and another begins, they can become confused. When readers become confused, they may quit reading entirely; when a confused reader is grading student work, the student may receive a poor grade.

To prevent confusion, professionally printed books and journals display paragraph separation carefully, by either indenting each new paragraph or by separating paragraphs with blank lines. The latter is called *block format*. For technical reports, it is advisable to indent and to insert a blank line after every paragraph. The extra line helps identify a new paragraph, and the indention avoids confusion with lines and blocks that may be skipped for exhibits. This combination may be called the *modified block format*.

Type size

For most uses, any type size between 9 and 12 point is generally acceptable [2]. Consequently, specific requirements should call for a suitable size in this range, probably 10 or 12 point type. If no font size is specified, students should select a large type size, such as 12 point, for ordinary printers and copiers. Professional printers can obtain good results with smaller sizes.

Font

The font is the graphical design of the *type face*. You select a font automatically when you open your editing program, and much of your document's appearance and spacing is driven by that font selection. Most editing programs provide users with a large number of type faces. However, you should create your class reports using either the Times New Roman or the Arial typeface. This is because these fonts are almost universally used; you can assume that most users will have these fonts installed on their computers, so you can share files with others and be confident that your writing and layout settings will reproduce consistently and correctly.

The biggest difference among fonts is the system of spacing and the presence of serifs. Serifs are the tiny lines at the ends of the long strokes. Proportional fonts have variable letter spaces depending on the size needed for the specific letter. Times New Roman (TNR) is a proportional serif font, as is **Garamond**, the font of this text. The Arial font used in many graphics is a proportional sans serif (*i.e.*, without serifs) font. Courier is a widely available non-proportional or fixed-space serif font. The "fixed-space" designation means that, as in mechanical typewriters, all the characters for this typeface fill the same amount of horizontal space; in Courier, the "i" character, for example, is as wide as the "m."

While the differences between typefaces are interesting for discussions of page design, these differences can become important if you email your reports to supervisors or clients who do not have your fonts installed on their systems. In such a situation, when there is a font mismatch, the other person's computer will substitute a typeface that may or may not work well with your document design. In some circumstances, a font mis-match can distort your report, which may negatively impress your reader. To avoid the risk of distortion in such situations, you should restrict your font selections to very widely-distributed typefaces such as Times New Roman and Arial.

Italic type

Italic type was developed to mimic handwriting. In most running text, it has been used for foreign words and phrases, and it is commonly used in reference lists for the titles of books and journals. In technical work, italic type has come to be used for math symbols in text, tables, and equations. Italic does stand out, but it is definitely slower to read [2]. Therefore, italic should be reserved only for those specific conventional applications.

Units

Most engineering institutions and organizations such as the AIChE, ASCE, ASME, and IEEE have adopted the Système Internationale (SI) units as the required sole or primary system for recording measurements. Consequently, SI should be used as the primary units for all undergraduate reports. Organizations such as ASCE and ASME also allow or even encourage SI units to be accompanied by United States Conventional Units (USCS). In practice, many industries and even some professional organizations such as ASHRAE still use USCS as the primary units, and they will surely continue to do so. Consequently, the students need to be familiar with both systems, and the technical author addressing the widest possible audience should try to include both systems when appropriate and allowed. For further guidance, probably the handiest and one of the best references on units is the online reference on constants, units, and uncertainty that has been developed and is maintained by the NIST [3]. Some specific guidelines are given below.

Primary units

Unless explicitly told otherwise, students should use SI as the primary system in all reports. In text, it is wise to accompany the SI values and units with USCS values and units in parentheses. An example is this: "1.05 m (3.44 ft or 41.3 in.)." When converting units, do not be tempted to use digits that imply a greater accuracy in the converted data than in the original. For example, "1.05 m (41.3 in.)" is acceptable, but "1 m (39.37 in.)" is not. Note the recommended special case of the period in the abbreviation "in." for inch. This period is unique in abbreviations for units; do not end other unit abbreviations with a period. When data in secondary units are desired in tables, use an adjacent column for the accompanying data.

Elementary SI units

The SI is based on seven elementary units of measurement that are summarized in Table 3.1 below. Note that an abbreviation is not capitalized unless it comes from a person's name. Also note that the name of the unit is not capitalized in text even when it does come from the name of a person.

Table 3.1. Elementary dimensions and units in the SI

Elementary Dimension	Name	Symbol
length	meter	m
mass	kilogram	kg
time	second	s
electric current	ampere	A
absolute temperature	kelvin	K
molar amount of substance	mole	mol
luminous intensity	candela	cd

Derived units

Numerous derived units have been defined based on the elementary units above. An obvious example is the mass density, which may be written as km/m^3 or $kg\ m^{-3}$ or $kg{\cdot}m^{-3}$. Note that division by a unit is indicated by the solidus, or slash dividing line, or a negative exponent. Multiplication of units is indicated by a blank space or a multiplication dot. Twenty-two derived units have been given names. Some of the more common derived units are given in Table 3.2. The NIST web reference or any equivalent resource will provide a full list of the derived units.

Table 3.2. Common combined dimensions and derived units in the SI

Combined Dimension	Name of Unit	Symbol	Elementary Definition	Other Definition
plane angle	radian	rad	$m{\cdot}m^{-1}$	none
solid angle	steradian	sr	$m^2 m^{-2}$	none
frequency	hertz	Hz	s^{-1}	none
force	newton	N	$m{\cdot}kg{\cdot}s^{-2}$	none
pressure, stress	pascal	Pa	$m^{-1}{\cdot}kg{\cdot}s^{-2}$	N/m^2
energy, work, quantity of heat	joule	J	$m^2{\cdot}kg{\cdot}s^{-2}$	$N{\cdot}m$
power, radiant flux	watt	W	$m^2{\cdot}kg{\cdot}s^{-3}$	J/s
electric charge	coulomb	C	none	$s{\cdot}A$
electric potential electromotive force	volt	V	$m^2{\cdot}kg{\cdot}s^{-3}{\cdot}A^{-1}$	W/A
capacitance	farad	F	$m^{-2}{\cdot}kg^{-1}{\cdot}s^4{\cdot}A^2$	C/V
electric resistance	ohm	W	$m^2{\cdot}kg{\cdot}s^{-3}{\cdot}A^{-2}$	V/A
electric conductance	siemens	S	$m^{-2}{\cdot}kg^{-1}{\cdot}s^3{\cdot}A^2$	A/V
magnetic flux	weber	Wb	$m^2{\cdot}kg{\cdot}s^{-2}{\cdot}A^{-1}$	$V{\cdot}s$
magnetic flux density	tesla	T	$kg{\cdot}s^{-2}{\cdot}A^{-1}$	Wb/m^2
inductance	henry	H	$m^2{\cdot}kg{\cdot}s^{-2}{\cdot}A^{-2}$	Wb/A
relative temperature*	degree Celsius	°C	$K - 273.15$	none

*Note that the relative temperature requires special attention. Please see the section on this topic.

Additional units

Several units are commonly used along with the SI, even though they are not on the recommended list. These include the obvious day, hour, and minute of time. Another obvious addition is the angular degree (*i.e.*, 180° = π rad). An important extra unit is the liter (*i.e.*, 1 L = .001 m³). Note that the capital L is the preferred symbol for the liter; it is used to avoid confusion with the numeral 1. The metric tonne (i.e., 1 tonne = 1000 kg) is also included. Note the spelling of *tonne* that can be used to minimize confusion with the conventional ton. While on the subject of tons, it should be mentioned that the ton of refrigeration is best identified as the U.S. Refrigeration Ton (*i.e.*, 1 USRT = 12000 BTU/hr).

Relative temperature

The SI does recognize the relative temperature, so the Celsius degree is defined according to the formula

$$t_C = T - 273.15$$

where 273.15 K is the conventional ice point. Differences or derivatives in either Celsius degrees or kelvins are identical. For example, a heat capacity could be written as kJ/kg·K or as kJ/kg·°C. Temperature differences are subject to confusion because a 100°C difference could be misinterpreted as a 373 K temperature instead of a 100 K difference. If differences and derivatives are written in kelvins, there is little or no chance of confusion; consequently, students should adopt this style. In a numerical example, consider 1.0 kg water, with a heat capacity of 4.2 kJ/kg·K, being heated from 25°C to 30°C. Then the temperature rise would be 5 K, and the heat required would be 21 kJ.

Conventional units

The conventional units are based on common usage and legal definitions in the USA. Most of the conversions between USCS and SI can be derived from the following exact definitions:

pound mass	1 lb$_m$	=	0.453 592 37 kg
foot	1 ft	=	0.3048 m
absolute temperature	1 R	=	1.8 K
BTU	1 BTU	=	1.055 055 852 62 kJ

To complete the usual set of conventional mechanical units, note that a pound mass weighs exactly one pound force under the standard gravitational acceleration of 9.806 65 m/s² making the lbf very nearly 4.448 222 N. To complete the usual set of thermal units, note that the exact definition above makes the BTU very nearly equal to 778.169 ft·lbf. For more details and values, a rather complete set of unit conversion factors is available at the NIST website.

Unit formats and abbreviations

Abbreviations should follow standard usage as illustrated in the tables presented earlier. Unit abbreviations are not math symbols; therefore, they are always in vertical text, properly called roman text. These abbreviations are never set in italics. Note that the single letter "s" not "sec" is preferred for the unit of seconds. Double line fractions such as $\frac{N}{m^2}$ are allowed for derived units, but they are conspicuous and awkward to enter as text. Note how that fraction disrupted the line spacing unnecessarily in this paragraph. Slash or shilling fractions such as N/m^2 are equally proper and simpler and do not disrupt the line spacing the way a double line fraction does. Obviously, slash fractions are much to be preferred. Periods are not used after abbreviations except in the special case of "in." for inch. Some editors omit the ° sign for the Celsius or Fahrenheit degree and simply write 25 C; however, the SI includes the degree Celsius and specifically includes the degree symbol, possibly to avoid confusion with the C for the coulomb of charge. Obviously, little confusion with the coulomb is likely in mechanical and chemical engineering literature if C is used for the Celsius temperature; however, organizations that have adopted the SI should probably follow the standard strictly and use °C. In any event, the degree symbol is never used in an absolute K or R temperature.

Other features specific to technical writing

Abbreviations

Typically, abbreviations should be avoided except for the conventional abbreviations for units that are discussed above. A long technical name can be replaced by an acronym formed from the initial letters of the name such as the acronym TNR used above to represent Times New Roman. The acronym should always be identified by a note in parentheses at the first appearance in a short report and probably at the first appearance in every chapter in a longer work. Even acronyms should be minimized, and they should always be identified in the text. It is always better to define your acronyms explicitly rather than to assume that the reader will correctly identify what it represents. For example, the acronym LSD may represent dangerous psychedelic drug, or it may represent the phrase "least significant digit." You should not give your reader the opportunity to associate your acronym with the wrong term.

Identifying equipment

It is helpful to the reader to describe the engineering or experimental equipment generically, rather than using a potentially unfamiliar commercial name or acronym. For critical instruments, you should follow up the generic description with a specific commercial identification. One way to do this neatly is to place the commercial description in parentheses. This style emphasizes that the commercial description is secondary. For example, an experiment might use a "mercury in glass thermometer (Omega, model ASTM-3C)." We call this style "generic (commercial)." Another style is to put the commercial description in a secondary subordinate clause or auxiliary phrase, such as describing a "mercury in glass thermometer, specifically an Omega, model ASTM-3C." For convenience, we refer to the latter style as "generic then commercial." The generic description explains the fundamental character of the instrumentation to the reader. The commercial identification might be helpful in convincing the reader that the instrumentation was adequate, and it would certainly be helpful in reproducing the experiment. The commercial identification may appear technical and well-informed, but it is essentially superficial. Instead, always emphasize a functional and fundamental description that will be meaningful to the typical technical reader.

References

References are crucial to scholarship and credibility. Professional style guides provide orderly methods to cite your references in the text and to provide complete publication information in a bibliography. There are many widely-used methods of citing and formatting reference entries, including the author-date format (the basis of APA reference style), and the citation sequence format (the basis of IEEE style). Both reference methods are widely used in the sciences, and professionals usually need to master both styles. In this book, for simplicity and consistency, we recommend that students insert references using a numbered-citation-sequence method such as that described by the *IEEE Editorial Style Guide*. Using the IEEE method, citations are inserted as an index number in square brackets, such as this: "[1]." In this method, your report's reference section uses the same numbering system, typing out the full publication information for each reference in the order of text citation. In this book, we use the IEEE method for our report illustrations, so that we can present a consistent approach to references. A wider discussion of citations and references is provided in Chapter 2.6.

Equation formats

Students have access to a variety of symbolic equation editors that make it possible to prepare clean and clear displays. For consistency in your reports, you should number your equations in sequence beginning with Equation 1. In the text of your reports, you should cite each equation using a specific equation number. The symbolic display of the equation should be centered on a separate line, as follows:

$$F = m\,a \tag{1}$$

The equation number should be enclosed in parentheses and placed with a consistent offset from the display; typically it is placed flush with the right margin. The variables in an equation should be defined on the first line following the symbolic display.

Equations and language

There are two ways to describe the grammatical status of an equation. First, an equation is an algebraic sentence or clause with its right hand side being the subject, left hand side being the object, and the equal sign—or inequality sign—being the verb. By this approach, an equation can be considered to be a clause that is part of a sentence or even as a separate sentence. By the second approach, an equation is a specific idea expressed in mathematical form, so an alternative definition is to think of it as a noun or noun phrase representing an idea. Equation 1 above is considered to be a noun phrase at the end of a sentence. The first interpretation considers the equation to be a clause. Therefore, one could write this compound sentence stating that special relativity relates the energy of a particle to its mass, and

$$E = mc^2 \tag{2}$$

Since an equation is an algebraic sentence, logic would indicate that it should end with terminal punctuation, such as a period or comma. However, the terminal punctuation after an equation has long been omitted and would now be considered hopelessly obsolete or just strange, so follow the convention and omit any end marks. A particular example is the period that seems to be needed but should be omitted at the end of an equation standing alone as a sentence or even after an equation ending a sentence of text such as this one:

$$dE = \delta Q - \delta W \tag{3}$$

For detailed information on formatting and editing equations see Chapter 2.10, which is dedicated to this topic.

Capitalizing exhibit names

Every individual writer needs a consistent practice for citing displays in print, and every course needs a policy that will support consistent and fair grading. In our undergraduate classes, our policy is to capitalize the first word of an exhibit citation, as if this were the exhibit's name. Thus, the first equation in a report would be cited as "Equation 1," the second table would be cited at Table 2," and the third figure would be cited as "Figure 3."

In contrast, less formal references to a display do not require capitalization. We might capitalize a table's formal citation by saying "Table 2 displays the calculated results," but we might later use lower-case characters when we speak of that table in an informal reference to "the table on the previous page."

Lists

Technical writing can be structured, simplified, and condensed by using lists. Two types of lists are used: "vertical" lists in column format, and lists in text style called "run-in" lists. A complex paragraph or section burdened by minutia can often be simplified and enhanced by moving the details to a list. See Chapter 2.13 for guidelines on the construction and editing of lists.

Parentheses

Parentheses are almost sure to disrupt the natural flow of a sentence, so parenthetical comments should be avoided. Unfortunately, the misuse of parentheses is widespread in technical writing. Indeed, they are the "duct tape" of technical writing, and they are warning signs of disorganization. Parentheses cannot be used to evade the rules of grammar or to shortcut the demands of clear writing, so novice writers should avoid this temptation and minimize their use of parentheses. In general, only the conventional uses of parentheses are recommended. Some common allowable examples include setting off citations and punctuating auxiliary explanatory notes and examples. Some transition is needed to identify the purpose of the parenthetical note, so it is good practice to start such notes and examples with an introduction. The conventional abbreviated introductions are both short and effective. So one can efficiently use either "*i.e.*" meaning "that is" to introduce an explanatory note or "*e.g.*" meaning "for example" to introduce an example. When one of these introductions to some information seems inappropriate, it may not be correct to consider the information as being merely parenthetical. Probably then the information is so important that it belongs in a sentence, or it is so trivial that it should be omitted. Some other acceptable and even mandatory uses peculiar to technical writing are listed in Table 3.3.

Table 3.3. Some acceptable or mandatory uses of parentheses in technical writing

Citing literature in (author, date) method	Providing alternative units as in 1.0 m (3.3 ft)
Punctuating indices in a list such as (1) or (a)	Punctuating equation numbers
Punctuating a source note in a table	Identifying a coordinate point as in (*x, y*)
Introducing an unfamiliar acronym as in laser Doppler anemometer (LDV)	Providing alternative commercial identification as in "dual beam LDV (TSI System 9000)"
Defining an open interval as in (*a, b*)	In algebraic formulas such as *b* (*a* + *c*)

Summary table

Table 3.4 summarizes the guidelines in this chapter.

Table 3.4. Guide to editing some features in text

Use a formal and legible font.	Avoid multiple fonts.
Use standard abbreviations.	Degree sign for temperature differences only.
Distinguish in. for inch.	Follow capitalization rules in abbreviations.
Use generic (commercial) or generic then commercial style of identification.	Avoid superficial identification.
Indent first line in semiblock paragraph.	Indent lists as needed.
SI units must be primary.	USCS may accompany in parentheses.
Sources must have standard citation.	References must have standard listing.
No informal citations.	No incomplete or informal listings.
Equations must be printed.	Use equation editor not text.
Every equation must be numbered.	Center equation on separate line.
Use lists to itemize and organize complex material.	Lists must be rhetorically parallel.
Avoid parenthetical discussions.	Minimize unnecessary parentheses.
Use parentheses when customary.	Use traditional introductions in parentheses (*e.g., i.e.*)
Proofread personally; do not rely only on the spell-checker.	Proofread backward sentence by sentence for grammar.
Proofread backward word by word for spelling and usage.	Avoid common usage errors (*e.g.,* it's for the possessive).

References

[1] P. A. Kolers, R. L. Duchnicky, and D. C. Furgerson, "Eye Movement Measurements for Readability of CRT Screens," *Human Factors*, vol. 23, p. 10, 1981.

[2] M. A. Tinker, *Legibility of Text*. Ames, Iowa: Iowa State University Press, 1963.

[3] B. Taylor and P. J. Mohr, "The NIST reference on constants, units and uncertainty," 1999.

CHAPTER 2.4

GENERAL GUIDELINES FOR EDITING EXHIBITS

Exhibits are crucial in technical writing. The exhibits explained here are graphs, illustrations, and other features that accompany the body of the text. One should become familiar with and use the editing guidelines, which are summarized in this chapter and further detailed in the subsequent chapters.

Cover page

For many reports, especially longer reports, a cover page is appropriate or required. If a cover is used, one should make it professional by following the ISO standard. Chapter 2.7 presents guidelines for cover pages and an example.

Graphs

Graphs are often the most important exhibits in a report, so they deserve special attention. The primary rule is to avoid clutter or illegible features in any graph. Every graph or other type of figure should have a unique number and a descriptive title. By convention, the title is displayed as a caption, and it is always located below the graph or illustration. On the graph, markers should be used to represent actual experimental data points. In contrast, distinct line types are used to identify model and literature curves. Unless you are told to do otherwise, it is best to use the "XY" or scatter graph template in your graphing program. As you scale your axes, be mindful of the (0, 0) point. If this origin point is physically significant, as with scaled variables, then one or both of the axes should begin at zero. The graph utility will automatically select (x_{min}, y_{min}) for the origin to maximize the resolution. For further guidelines and detailed instructions, please see Chapter 2.8, which is dedicated to graphs.

Illustrations

In addition to graphs, other figures, such as schematics of apparatus, are often included in technical reports. You should keep these illustrations simple, legible, and uncluttered. For further guidelines and detailed instructions, please see Chapter 2.9, which is dedicated to illustrations.

Equations

These have been discussed as text elements, so please refer to that topic as discussed earlier in Chapter 2.3 for an overview. Further, more detailed guidelines are presented in Chapter 2.10.

Spreadsheets

Since they are widely used for processing experimental data, spreadsheets are commonly required attachments for undergraduate lab reports. They are used in monitoring one's data and calculations. When it is submitted with a report, a spreadsheet is usually included as an attachment, which is a single item appendix, to a report. Chapter 2.11 presents guidelines on preparing the spreadsheet file to receive and process data. Entire spreadsheets or even large blocks are usually too expansive and informal to be included in widely published professional reports. In contrast, a block from a spreadsheet or even an entire concise spreadsheet may be attached to a report that is designed for limited circulation. As noted above, the most pertinent data must often be extracted and presented in a separate and concise table, which is then integrated into the body of the report. When this happens, be sure that this attached spreadsheet has a unique number and name that are centered and prominent on the page such as "Attachment 1. Experimental Spreadsheet."

Printing or attaching spreadsheets

A spreadsheet may be printed out separately and appended to a report. Preferably, a picture of a suitable and pertinent block can be attached to a report. When printing out a spreadsheet, pay close attention to the page design. If several pages are required, be sure that each individual page is coherent and meaningful. A printout is sometimes acceptable, but it is rather informal. Creating a concise spreadsheet attachment is the preferred presentation. If a particular calculation in the spreadsheet is important, unusual, or complicated enough to require special description and explanation, it may be appropriate to print out a small block of cell formulas to illustrate the calculation in a separate attachment with a separate identification. Never print a huge block of formulas along with less relevant information such as numerical data, labels, and so forth. Print only a pertinent, representative block implementing the critical calculation or algorithm such as a numerical integration. It is good practice to describe the calculation or algorithm represented by the block of cells in an accompanying paragraph of text.

Tables

Integrating a concise, well-organized table into the body of the report is much preferable to appending and citing a longer and typically less formal spreadsheet. The unique number and descriptive title of the table should be centered above the block of data.

Details in tables

In a table, be sure to label the columns and include the units as needed. Since the primary units are typically SI, the units in the table should typically be SI. An uncommon exception would be a specific requirement to report raw data measured in conventional or special units. When this happens, you should accompany such data with a column showing the raw values converted to SI. Eliminate insignificant trailing digits when you do such a conversion. For example, use 3.34 m/sec rather than 3.342 m/sec for a speed measured to about 1% accuracy. Consider using Courier typeface to make the numerals more legible. The following example, Table 1.a, shows minimally acceptable Times New Roman typeface in the table, and Table 1.b shows better practice by using Courier typeface for table cell entries. For more guidance on table preparation, see Chapter 2.12.

Table 1.a. Velocity Profile Data

position mm	air speed m/s
0.15	1.23
0.2	2.01
0.4	4.52

Table 1.b. Velocity Profile Data from Pitot Probe Scan

position mm	air speed m/s
0.15	1.23
0.2	2.01
0.4	4.52

Lists

Lists have been discussed thoroughly as a text element, so please see that topic as discussed in Chapter 2.3 and the detailed guidelines in Chapter 2.13.

Appendices

Use appendices or other similar attachments for long mathematical derivations or calculations or for presenting other auxiliary work. Appendices may be identified by a letter (*e.g.*, Appendix A) or with a Roman numeral (*e.g.*, Appendix I). Give every appendix a unique number and a descriptive title (*e.g.*, Appendix A. Sample Energy Balance Calculations). Every appendix must be cited in the text; for consistency in undergraduate reports, assume that a reference to a particular appendix by its unique identification number is analogous to referring to a person or place by a proper name. Therefore, one should capitalize the word, as in these example citations: "as shown in Appendix 1" or "as detailed in Appendix A." However, the word is not capitalized in "as shown in the appendix," as this is not a formal citation by name. Remember that appendices are still part of the report and that all features and details must maintain high standards of technical accuracy, scholarship, and production quality.

CHAPTER 2.5

CONSOLIDATED EDITING CHECKLIST

The following checklist is largely consolidated from the pertinent chapters of this text. Refer to the subject section for further guidance. This checklist is a general and generic guideline only. Students are responsible for a neat and professional presentation that is legible and cogent. Every possible error cannot be listed here, but many common problems are addressed.

Table 5.1. Consolidated editing guideline

General aspects

Report is prepared in correct format, *e.g.* itemized or memorandum.	Cover page, abstract, and table of contents included if required.
Report is dated and signed.	All required or necessary items and topics are addressed.
The writing style is effective, concise, scientific, and efficient.	The report is complete and well organized.
Padded text and redundant comments avoided with no excessive length.	Jargon or colloquial writing avoided.
Conjectures and vague or subjective arguments are avoided.	All conclusions are justified on scientific and empirical basis.
Text is grammatical and proofread, not merely spell checked.	The report is securely and safely bound.
Introduction defines project and answers the questions, WWWWH.	Closure is complete, concise, and effective.
REFERENCES	
Citing and listing of references is conventional and scholarly.	Page and text formats are conventional, correct, and consistent.
EXHIBITS AND ATTACHMENTS	
All exhibits and attachments cited are included.	All exhibits and attachments included are cited.
All exhibits and attachments are of professional quality.	Named exhibits (*e.g.* Table 1 not "the table") are capitalized.
UNITS AND DIGITS DISPLAYED	
SI units are primary (USCS may accompany).	No insignificant digits in text or tables.
STATISTICS	
Adequate attention given any statistics and regression analysis.	R-squared and alpha risk addressed for every regression model.
PAGE, SECTION, AND PARAGRAPH FORMATS	
Margins are uniform on top and bottom and on both sides, usually one inch.	No wide bottom margins appear especially at exhibits.

Table 5.1. Consolidated editing guideline *(cont'd)*

Section headings are used and accurately identify the role of the text.	Text in a section addresses the topic completely without misplaced remarks.
Modified block paragraph or other required format used consistently.	Paragraphs throughout text have clear objectives and address requirements.
PARAGRAPH STYLE	
Paragraph topic sentence is present.	Paragraph topic is developed.
Extra line is skipped between paragraphs.	
SENTENCE STYLE	
A half line is skipped between every line of text.	Conventional sentence grammar is used.
Incomplete fragments or run-on sentences are avoided.	Simple sentences predominate with only isolated compound or complex sentences.
The text is impersonal.	First person is avoided or minimized.
Active voice used judiciously.	Parallel structure in list or series (*e.g.*, imperative sentences in listed procedure).
Common typo errors (*e.g.* "too" for "to", "it's" for "its") avoided.	Common technical writing faults are avoided.

Parentheses in technical writing

Citing literature in (author, date) method.	Providing alternative units as in 1.0 m (3.3 ft).
Punctuating indices in a list such as (1) or (a).	Punctuating equation numbers.
Punctuating a source note in a table.	Identifying a coordinate point as in (x, y).
Introducing an unfamiliar acronym as in laser Doppler anemometer (LDV).	Providing alternative commercial identification as in "dual beam LDV (TSI System 9000)."
Defining an open interval as in (a, b).	In algebraic formulas such as b (a + c).

Table 5.1. Consolidated editing guideline *(cont'd)*

Features in text

Use a formal and legible font.	Avoid multiple fonts.
Use standard abbreviations.	Degree sign for temperature differences only.
Distinguish in. for inch.	Follow capitalization rules in abbreviations.
Use generic (commercial) or "generic then commercial" style of identification.	Avoid superficial identification.
Indent first line in modified block paragraph.	Indent lists as needed.
SI units must be primary.	USCS may accompany in parentheses.
Sources must have standard citation.	References must have standard listing.
No informal citations.	No incomplete or informal listings.
Equations must be printed.	Use equation editor not text.
Every equation must be numbered.	Center equation on separate line.
Use lists to itemize and organize complex material.	List must be rhetorically parallel.
Avoid parenthetical discussions.	Minimize unnecessary parentheses.
Use parentheses when customary.	Use traditional introductions in parentheses (*e.g., i.e.*)
Proofread manually; don't rely only on spell checker and grammar checker.	Proofread backwards sentence by sentence for grammar.
Proofread backwards word by word for spelling and usage.	Avoid common usage errors (*e.g.*, it's for the possessive).

Citing and listing references

Recognized citation style is used.	One citation style is used throughout except for unusual circumstance of occasionally citing numerous works.
All sources cited are listed in the reference section.	All references listed in the reference section are cited in the text.
Reference listing has complete bibliographical information.	Consistent capitalization system used in all listings of references.
Listing is in proper order with proper punctuation.	Italics are used for books and journal titles and quotes used around titles of reports and other shorter works.

Table 5.1. Consolidated editing guideline *(cont'd)*

Graphs for experimental engineering reports

Design is neat, uncluttered, legible, and effective.	Unique number is assigned and descriptive caption is given.
Caption always below graph in written report.	Consistent capitalization system used in all captions.
XY graph, not line chart, is almost always used.	Graph box included only with discretion and then consistently used in one report.
No reliance on color or subtle gray scales.	No shading of graph area.
Graph is cited in text.	Any cited graph is included.
Graph is inserted into text at first feasible location when possible or on separate page when necessary.	Graph is attached (*i.e.* as a one item appendix) to extended abstract only.
Full page graph on separate page in landscape orientation has caption to right.	Separately printed graph is pasted onto page or has machine printed caption pasted on to graph page.
Zero point is purposefully and appropriately included or excluded in axes.	Secondary axis is included when necessary.
Axes are not so compressed or span is not so great that variation in data is not apparent.	Special care taken not to obscure scaling relations between normalized variables.
Axes are titled, with dimensions if existing.	Markers are strictly reserved for discrete data.
Data is presented in scatter plot or connected profile of markers as appropriate.	Distinct line types (*i.e.* not markers) are used to identity continuous data or model curves.
The series in a single series graph is identified only in caption.	All series are identified in legend, in notes on graph, or in caption.

Illustrations and drawings

Figure is neat, uncluttered, legible, and effective.	Unique number is assigned and descriptive caption is given.
Caption always below figure in written report.	Consistent capitalization system used in all captions.
Photograph used only with discretion and usually accompanied by schematic.	Usually a schematic or simplified drawing is preferred.

Table 5.1. Consolidated editing guideline *(cont'd)*

Equations

Equation has logical grammatical setting.	Equation is neat, uncluttered, legible, and correct.
Equation is generated with Equation Editor not a line of text.	Equation is centered on separate line with unique number at right margin.
Equation is proper size, especially for projection slide.	Equation box is sufficiently wide to avoid truncation.
Templates from menu are used for parentheses, not symbols typed from keyboard.	Math symbols are in Italics with functions, numbers, and notes in regular type.

Spreadsheets

Overall design and detailed format are neat, uncluttered, legible, and effective.	Unique number is assigned and descriptive title is given in heading.
Spreadsheet is presented as an attachment cited in text.	Concise block of pertinent data is also extracted and presented in a table.
Title heading is centered and prominent with type size at least 10 point.	Spreadsheet is structured with heading, preliminaries, summary, and data blocks.
Heading identifies purpose and history of spreadsheet.	Unique data, constants, and parameters are placed in the preliminary section.
Summary section is near top of spreadsheet.	Routine recurring data are in a well-organized separate section.
Consistent numerical format is used in a related block or blocks of data.	Units are identified for every dimensional data block, typically in column heading.
Recurring data are identified, usually with column header.	Unique data are identified, usually with identifying phrase.
Printed margins are adequate and at least 1 inch all around.	Multiple page printout is carefully designed especially in continuing pages.
Block of formulas presented in separate attachment if required.	Optional concise summary regression block prepared if appropriate.

Tables

Table is cited in text.	Table is concise, neat, uncluttered, legible, and internally consistent.
Table has unique number and descriptive title in heading.	Table is centered horizontally on the page.
No insignificant digits are displayed.	Data are identified, usually in column headers, units are given unless data are nondimensional.
Table is limited to one page.	Numbers are properly aligned.
Consistent numerical display format used except as limited by significant digits.	Scientific notation used for very large or very small magnitudes or to carefully identify significant digits.
Appropriate text format, usually centered, is used for headers and notes.	Source and descriptive notes and footnotes are provided.

Table 5.1. Consolidated editing guideline *(cont'd)*

Lists

List used when appropriate and helpful to simplify and enhance presentation.	List not used for incommensurate items.
Run-in list has no extraneous punctuation.	Vertical list can have bullets, numbers, letters, or no indices as appropriate.
Use multiple column list for brief items.	Optional right indention of vertical list is helpful.
List must be rhetorically parallel.	Consistent capitalization system used in all lists, usually initial capital only.

CHAPTER 2.6

STYLE GUIDES AND REFERENCES

What are style guides?

A style guide is a book of specifications used by publishers and printers to ensure that books and journal articles look professional, that they have a consistent layout and appearance, and that they accommodate their readers' particular interests. Because professional organizations commonly have publishing components, most professional organizations have developed style guides of their own. Such organizations include, but are surely not limited to, the American Institute of Physics (AIP), the American Society of Mechanical Engineering (ASME), and the American Institute of Chemical Engineers (AIChE).

Among book-length style guides, the most widely used style guides include *The Publication Manual of the American Psychological Association* (APA), The Institute of Electrical and Electronics Engineers (IEEE) *IEEE Standards Style Manual*, and *The Modern Language Association Style Guide* (MLA). The Turabian *Manual for Writers of Research Papers, Theses and Dissertations* was produced by the University of Chicago Press to govern the production of graduate theses, and it has become widely accepted as a style guide for academic writing. The University of Chicago Press also publishes the authoritative *Chicago Manual of Style*.

In addition to specifying document design and page layout, style guides offer more or less detailed discussions of ethics in research, appropriate approaches to research documentation, data preservation, conflicts of interest, and acknowledging the contributions of colleagues, subordinates and other researchers.

Most professional journals provide brochure-size style guides, which are usually listed as "author information." These guides only briefly define the layout and format requirements of the journal, although they typically provide software templates that encode the journal's print standards. Beyond this, author information guides usually provide considerable information about copyright and ownership of your submitted documents, and they will usually disclose the review criteria and procedures used by the journal's review editors.

Most students will have limited direct contact with professional style guides. However, most instructors will provide specifications for the reports in their courses, and these specifications can be treated as a simplified style guide. Usually an instructor's classroom style guide will be based on the guides provided by professional organizations—to familiarize students with professional standards—but with adjustments to facilitate efficient grading. In most cases, students will be aware of the instructor's style guide only in the format that is defined for reference citations and for the list of works cited. The remainder of this chapter focuses on these.

Citations and References

Citations are the notes in the text that identify a source of information from the technical literature or other resource. The references are the listing of the sources. One should cite all references in the text and list the bibliographical references at the end of the report.

Why are references needed?

When writing a lab report or other technical document, there are several reasons for using references. First, we use references to establish credibility. In engineering and scientific writing, the author's own opinion is typically not enough to back up a claim. Finding sources that support stated conclusions can help to provide the authority needed. Secondly, references can put the experimental objectives into a meaningful context. Finding relevant background sources lends perspective to the study at hand and can provide motivation for the current study and that can satisfy those readers who would like to delve more deeply into a topic. Finally, and related to context, we use references to distinguish work that we have done from work that was performed by others.

What makes a good reference?

First, a research reference must be credible. Sources such as textbooks, technical handbooks, and peer-reviewed scientific journal articles are three examples of credible sources. Peer review refers to the process through which manuscripts are accepted by journals for publication; the "peers," other academics in the field, are enlisted by journal editors to anonymously review submitted manuscripts in a multi-step process designed to ensure quality and accuracy in the scientific literature. Certain online encyclopedia sources like Wikipedia can be good starting points for research, but they are not typically acceptable as references because they may not undergo extensive and transparent review or editing processes that are used by academic journals.

References must also be relevant to the topic at hand. For an undergraduate report, journal articles that are focused on a minute aspect of a process may not be as helpful as a review article that gives a more general overview of the current state of the art. And of course, a reference should be published, so that it is accessible to the reader. References such as personal communications or unpublished lecture notes should be used sparingly, if at all.

Students often ask whether they should cite things that they just "know." Our answer narrowly addresses your data--your measured values and your analytical results. All the data in your report came from somewhere: either you directly measured and calculated a value, or you used values that were measured and calculated by others. You need to cite sources and keep good records in order to keep this distinction clear. First, keeping written records of experimental results can sharply define your work, which will help to establish the novelty of your work, and this is useful when you apply for a research grant, fellowship, or patent. Second, proper citations for information that came from another source will lend more credibility to you as an author. After all, a reader is more likely to trust your claims if you can show that other published studies repeat this same claim.

How can I find credible, relevant references?

The first step to take in finding references is to look through your textbooks. Most textbooks have withstood the test of time and undergone multiple cycles of editing and revision; therefore, even if these sources are not the most current, they are often the most reliable and easily accessible for undergraduates. You should also look at the reference sections within your textbooks, as that can be a good starting point for your own research on similar topics.

Another modern and convenient research tool can be found online: article databases. If you are enrolled at a college or university, chances are that through your library website, you will have access to databases of technical articles. Some of the most common databases for technical literature are Web of Science, SciFinder, and Science Direct. These databases are created by academic professionals and only include peer-reviewed articles from reputable scholarly journals. Databases also exist for specific disciplines, and even for sub-topics within that discipline. Searches by keyword, date, and even the number of times cited can help you find the most relevant sources for your topic. What's more, students can usually just click on the search results to access the desired articles.

Using search engines such as Google or Bing to find references may seem convenient, but this method is not ideal for a few reasons. First, the search engine will search the entire Web, not just the scientific literature. Consequently, you must spend some time to sort through the results. Second, the sheer number of search results can reach into the thousands, requiring many hours to review. One popular option that splits the difference between scientific databases and search engines is the free scholarly search engine Google Scholar, which "crawls" over the full text of millions of published articles--even those behind the paywall. One helpful feature is that it lists results according to relevance to your key words, rather than in reverse chronological order, which is often what databases do. One potential drawback of Google Scholar is that the search process used by Google lacks transparency, and some pseudo-scientific, non-peer-reviewed articles may find their way into search results. But when used with caution, Google Scholar can help provide appropriate references for a student research paper or lab report.

Citation Style

Like most elements of professional document styles, citation and reference styles were developed to support the particular needs of individual disciplines. The author-page number citation style of the Modern Language Association helps readers to keep track of page numbers for different editions of book-length prose or poetic works, but this is not used in the sciences, which tend to rely on short journal articles. Citation by footnote is widely practiced in legal studies and the humanities, where relevant arguments are best introduced at the moment when the reader encounters a relevant point. But this method has fallen out of use in science and technology.

The numerical sequence format is used by numerous professional societies, such as IEEE, ASME, and ACS, and it is common in engineering publications. In this format, a reference is cited by its sequential list number. The number is usually typed in brackets, *e.g.*, [1] or Reference [1] or is superscripted. At the end of a sentence, the closing bracket precedes the period. In the References section, numbered references are listed in order of appearance, making it easy for readers to identify and group clusters of related works.

The author-date method is also called the Harvard style and is used in APA format, among others. This method creates reference lists that cluster the works of individual authors, making it easy for readers to identify authorities in a particular domain and to prioritize the works by date. The author-date format is most widely used in the social sciences and humanities, but it is common in science and engineering as well. In this reference style, the source is identified with the author's last name and the year of publication, *for instance*, (Smith, 2012) or Smith (2012). As just exemplified, the name of the author may be part of the sentence, or it may be parenthetical, depending on the context. When a given author has produced several publications in a single year, these are distinguished with small letters, such as Smith (2012a) and Smith (2012b) in the citation and listing.

Reference management programs such as EndNote and BibTek can be used to search for sources in the literature, create reference libraries, and automatically format in-text citations and references to comply with most commonly used reference styles. With these types of programs, students can avoid manually entering citations and can also painlessly add or change references without having to change all of the numbering. EndNote can be particularly useful when writing a longer research report or design report that requires a greater number of references. An example of an EndNote screen with the IEEE style selected is shown below.

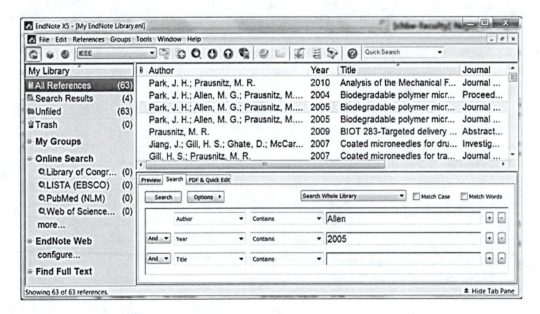

Figure 6.1. This screenshot from EndNote X5 highlights the ability to perform online searches of multiple databases and to create a searchable library of references. It also allows students to automatically format references using a particular style. (Here, IEEE style is selected in the upper left of the screen.)

With EndNote, you are able to select virtually any style that is required by your instructor or by a publication. The next figure shows an example of the dialog box wherein you may select a reference style.

Figure 6.2. Shows the "Select a Style" dialog box from EndNote, where the citation style preferred by your instructor can easily be chosen. Note that in addition to the almost 500 styles built into EndNote, others can easily be downloaded from endnote.com, which offers over 6000 styles.

A number of other tools are available for students who seek assistance in formatting a list of references. These include OttoBib, eTurabian, citethisforme.com, and KnightCite, as well as the built-in reference tool of Microsoft Word. OttoBib, for example, can generate a complete reference entry from a book's ISBN number; however, this is not effective for journal articles. KnightCite, like the Microsoft tool, can generate acceptably formatted reference entries, but it requires the user to manually create these by entering the author(s), title, journal volume, and so forth. Citethisforme.com can import an existing bibliography and manage the format; it, like eTurabian, is integrated with a search tool that extracts reference information from electronically located sources. These tools, like BibTex and Endnote, are integrated with search tools that can greatly improve an author's productivity.

Listings

A complete list of the bibliographical references cited in the text, with every entry in an approved format, should follow the main text of the report. Please note that various publishers and organizations will differ in style and content of the listings. Nevertheless, your listing must be descriptive, definitive, and consistent. An adequately descriptive listing identifies the source and the nature of the source. The reader needs to know not only the author, but the subject and the publisher of the reference as well. A knowledgeable reader will appreciate recognized authorities and distinguish refereed from less reliable non-refereed or commercial literature. A definitive listing is complete enough that the reader could actually find the source independently. Finally, the listings should be consistent both among themselves and with the guidelines of the editor or, in academe, the instructor. Remember, incomplete or inconsistent listings draw unwanted attention and raise questions about the writer's attention to content and details. Always record a complete, not minimal, description of your sources during your literature research so that you can generate later whatever listing is required. Alternately, programs such as EndNote can be used to locate and keep track of sources and can also automatically generate reference lists that are formatted to a particular style. The table below shows the reference styles that are most common in engineering and other technical fields.

Table 6.1: Common reference styles in scientific and technical fields

Field	Common Reference Styles
Aerospace Engineering	ASME (American Society of Mechanical Engineers); ASCE (American Society of Civil Engineers)
Biomedical Engineering	ACS (American Chemical Society); AMA (American Medical Association)
Chemical Engineering	ACS
Civil Engineering	ASCE
Electrical and Computer Engineering	IEEE (Institute of Electrical and Electronics Engineers)
Mechanical Engineering	ASME; IEEE
Computer Science	IEEE
Medicine	AMA
Physics	AIP (American Institute of Physics)
Psychology	APA (American Psychological Association)

Unless some specific contrary rules are given, the student or engineer can safely use the style listed in this section, which follows the IEEE format used in many scientific journals. The complete IEEE format guide can be found online. Note that in most reference styles, including IEEE, journal titles are abbreviated. The websites for professional societies such as ASME, ACS, and IEEE typically list the accepted abbreviations for their most commonly cited journals. Even if another style is eventually required, the student is safe to begin work with the system described here in mind. Since this system is complete and consistent, references that fit it can easily be altered to satisfy any reasonable alternative rules. There is only minimal consistency among editors about details such as the order of the parts and the separators (*i.e.*, commas or periods). Consequently, always understand and follow the particular style that may be required. The list itself may be single spaced with a blank line between each entry. There is even a difference in the style of indenting the list. Some editors indent the first line of an entry while others print the first line flush and indent the subsequent lines. The IEEE format presented here does not require any indentation.

Note that publishers and conferences often provide authors with editing templates that have reference formats built in (especially LaTeX templates). When a template is not available, most publishers will offer a guide to reference formatting, similar to what is offered in this section.

Some general forms and several specific examples for various types of literature sources follow.

Journal articles

Journals are periodical technical or professional publications such as the ASME journals in heat transfer, fluid mechanics, and other fields. Journals usually publish the results of significant and original research. Usually most technical references are to journals, and the listing of a journal reference is about the most typical type. The general form is as follows:

Initials and last name of first author, followed by initials and last names of any other authors, "Name of Article," Abbreviated *Title of the Journal in Italic Type*, vol. x, no. x, pp. xxx-xxx, Abbrev month, year.

A typical example follows:

B.W. Wepfer and C.L Haynes, "Enhancing the Performance Evaluation and Process Design of a Commercial Grade Solid Oxide Fuel Cell via Energy Concepts," J. Energy Resour. Technol., vol. 124, no.2, pp. 95-104, June 2002.

Published conference proceedings

These compilations are the published records of the papers presented at the regular meetings of a technical society or at a special technical conference. This type is similar to a journal publication except the title of the transaction or proceeding replaces the title of the journal as follows:

Transactions:

Initials and last name of first author, followed by initials and last names of any other authors, "Name of paper," In *Title of Transactions*, vol. x, pt. x, pp. xxx-xxx, year,.

N. Kumbhat, P.M Raj, R.V Pucha, J. Tsai, S. Atmur, E. Bongio, S.K. Sitaraman, and R. Tummala, R., "Novel ceramic composite substrate materials for high-density and high reliability packaging," in *IEEE Transactions on Advanced Packaging*, vol.30, pt.4, pp 641-653, 2007.

Published Conference Proceedings:

Initials and last name of first author, followed by initials and last names of any other authors, "Name of paper," In *Unabbreviated Name of Conference*, City of Conf., Abbrev. State, year, pp. xxx-xxx.

N. Hotz, "Solar-powered reformed methanol fuel cell system," In *ASME 10th International Conference on Fuel Cell Science, Engineering and Technology*, San Diego, CA, 2012, pp. 81-89

Books

Books are longer works usually edited and distributed by a publisher. The publisher is assumed to have some independence from the author and some supervision of the contents. Technical books usually are only secondary references based on primary references that are usually journal publications. For publication-worthy papers, the best practice is to review and cite the primary reference. However, for student reports, it is permissible to cite books, including textbooks. Always distinguish books from long reports, which have no independent publisher. You should identify the page range where the pertinent information is located. Always identify the publisher and the address of the publisher as follows:

Initials and last names of all authors, *Title of the Book in Italic Type*, xth ed., City of publication: Publisher, Year of publication, pp. xxx-xxx.

A typical example for a book follows:
F. M White., *Viscous Fluid Flow*, 3rd ed., New York: McGraw-Hill, 2005, pp. 74–76.

Obviously, the edition number is not needed on a first edition unless a second or later edition is known to exist.

Research reports with limited circulation

This category includes all internal reports and reports with limited or restricted circulation. Short reports may resemble journal articles, and longer reports may physically resemble books. Nevertheless, avoid assigning them the same status since the author of a report is essentially the publisher, and no independent review or supervision can be assumed. This listing is similar to that for a book, except the title is placed in quotes instead of being printed in italics. In addition, include any identifying number for the report and identify the sponsor if known and appropriate. Include the institutional or corporate affiliation of the authors, the entity that in effect serves as the publisher of the report. An example of interest to a small group of undergraduate lab students follows:

C. C. Pascual and S. M. Jeter, 1998, "Measurement of Heat Leak from the Copper Cylinder Used in Convection Heat Transfer Experiment," The George W. Woodruff School of Mechanical Engineering, Georgia Institute of Technology, Atlanta, GA, 23 November 1998.

This general form for a research report can be modified as necessary and used for a range of narrowly distributed writings such as calibration reports, student papers, student theses, and course manuals and materials.

Commercial literature

In engineering and especially in experimental engineering and design, references to commercial literature such as specifications, catalogues, and operating manuals are common. A modified form of the report style should suffice. Look for a copyright note to find a publication date. Assume a corporate author unless a personal author happens to be identified, as in this example:

SKF USA, 2010, "MRC Engineering Handbook," SKF USA, Inc.., Lansdale, PA, pp. 53-60.

Personal communications

Because personal communications such as a conversation or message usually cannot be documented in print, using them as sources should be avoided. Nevertheless, situations arise where such sources must be used, and citations will be necessary. Examples include highly specialized data, an eyewitness account, or a personal observation. The engineering investigation of an accident or failure is one example when such sources may be necessary. Be sure to record any such source with an entry in your research notebook. Use the following general format, always identifying the type of communication, such as letter, lecture, conversation, phone conversation, e-mail, and so forth:

> Last name of correspondent or communicator, followed by initials, year, type of communication, such as letter, lecture, e-mail message, phone conversation, or private conversation, place conducted or initiated, specific date.

An example that might refer to specific information needed in an undergraduate report follows:

> Donnell, J., 2013, private conversation, Atlanta, GA, 12 August, 2014.

Lectures and Presentations

For student reports, it may be necessary to cite notes from a lecture or presentation. Again, it is always better to cite the primary reference for such material. But in certain circumstances, citing lecture notes is permissible.

> Initials and last name of lecturer, "Title of lecture/presentation," Presented in Number: name of course, *Name of Institution*, Full date (include month, date, and year)

> P. Ludovice, "Mathematical modeling," Presented in ChBE 4200: Unit Operations, *Georgia Tech*, May15, 2011.

Film or video

Occasionally a video or film must be cited. Identify any principal performer or presenter as in a technical lecture. Use the following form and identify the technical format, such as DVD, when possible. If viewed online, give the URL and date observed as in the following:

> Last name and first initials of principal (if any), year produced or uploaded, Title in quotes. URL (if viewed online) (Date accessed).

An example of interest to undergrad lab students follows:

> "Erasing with heat," 2015, Brightcove Video Cloud. http://www.nsf.gov/news/mmg (Accessed March 18, 2015).

Patents

Patents are cited in roughly the same way as journal articles, as follows:

> Initials and last name of first author, "Title of patent," U.S. Patent x xxx xxx, Abbrev. Month day, year.

S. Banerjee, "Method for indirect detection of nonelectrolytes in liquid chromatography," U.S. Patent 4734377 A, Mar 29, 1988

Internet sources

Finding and reading articles and books online is perhaps the easiest and most convenient method of locating relevant sources for a report. Many, if not most, technical journals now have an online version as well as a print version. Another prominent trend is towards online-only journals and textbooks.

For journals and books that are both printed and published online, the normal reference style should be used. However, the reference should note that you have accessed the online version of the article. This distinction is important because it is possible that the online version may have been updated more recently, and in ways that the print version may not have been.

Online journals that are based on print editions

You should use the normal style for a journal article, but include the notation "Online" in brackets after the abbreviated journal title. After the publication date, you should also include the URL and the date accessed:

B.W. Wepfer and C.L Haynes, "Enhancing the Performance Evaluation and Process Design of a Commercial Grade Solid Oxide Fuel Cell via Energy Concepts," J. Energy Resour. Technol. [Online], vol. 124, no.2, pp. 95-104, June 2002. http://energyresources.asmedigitalcollection.asme.org/ article.aspx?articleID=1414110 (accessed April 22, 2015).

Online-only journals have become more and more popular in recent years because of changes in the way most faculty, students, and others choose to access information. Another factor is that many academic libraries are transforming themselves from repositories of printed information to dynamic resources for online research, student and faculty collaboration, and communication. As such, there is simply not room on the shelves anymore for printed copies of every academic journal. Moreover, online-only journals may offer added features such as links to YouTube videos, spoken commentaries, and the like. Thus, peer-reviewed articles in online journals are a worthy addition to your reference list.

Journals that are published only online

For online-only journals, you should follow the normal citation style for print journals, but be sure to include the notation [Online] after the article title. You must also include the direct URL of the article and the date that you accessed it.

If the volume and number are not listed, you may include other identifying information instead such as the article number or digital object identifier (DOI).

Journal articles that are published online in advance of print issues

Often, journals will publish edited, peer-reviewed articles online well in advance of the print issue of a publication. These are usually identical to the printed version, but they often lack page numbers and volume numbers. When listing this type of source, include the designation "Online early access" in brackets after the abbreviated journal title. You may also include the DOI. Finally, you should list the direct URL for the article, as well as the date you accessed it.

> M. Baruch, J. Pander, III, J. White, A. Bocarsly. "Mechanistic Insights into the Reduction of CO2on Tin Electrodes using in Situ ATR-IR Spectroscopy." ACS Catalysis [Online early access].Publication Date (Web): April 13, 2015. DOI http://pubs.acs.org/doi/pdfplus/10.1021/ acscatal.5b00402. DOI:10.1021/ acscatal.5b00402 (accessed April 22, 2015).

Websites

While finding sources on the Web is quick and convenient, students must use caution in determining which sites constitute credible, authoritative sources. In most science and engineering classes, citing popular periodicals such as *Time* or *Newsweek* is not encouraged, as these are not academic peer-reviewed sources. Additionally, one of the most popular reference websites, *Wikipedia*, is generally not considered a credible source for a research paper. Citing Wikipedia, About.com, or other such online encyclopedic sources would be similar to citing the Encyclopedia Britannica: acceptable for a middle-school essay, but far too general for a college report. Wikipedia can be an excellent starting point in researching a topic, but after reading the entry, your next step should be to look up the references that are listed at the bottom of the page.

Still, there may be cases where it is appropriate to cite a website: perhaps you are looking up literature values or manufacturer data that are only listed online. Academic websites, such as the website of a research group or professor at an accredited engineering program, may also provide information that can be useful and credible in a student report. In these cases, you may use the format below.

General or academic website

> Author's first initial and last name, if known (and/or, the organization's or company's name).*Title of web page.* [Online] Direct URL (Accessed Month day, year).

> SPIE. *'Lab-on-a-chip' enables desktop instruction in fluid mechanics.* [Online] http://spie.org/x113148.xml#'Lab-on-a-chip' (Accessed April 28, 2015).

Plagiarism

One of the most important considerations in writing any report is to avoid plagiarism. Plagiarism is defined as using content or ideas generated by others without permission, or without giving proper credit to the original authors. For students and professionals alike, any instance of plagiarism can cast doubt upon the author's entire body of work, past and present; it is an egregious form of academic dishonesty. You may be familiar with cases of plagiarism covered by the media in which politicians or academics copy content from another source without attribution. Whether or not these people intended to commit acts of academic dishonesty is practically irrelevant. What matters is that they borrowed words or ideas that were not their own, and they did not give credit to the original source. Over the past 15-20 years, plagiarism has become more rampant because online sources have made copying easier, and the temptation to plagiarize has become greater. Copying and pasting a sentence or two from a website without citation may seem harmless, but such actions constitute unethical behavior and will be treated seriously by instructors and editors. Consequences for plagiarism at colleges and universities range from a failing grade on the report to expulsion from school. Therefore, this matter merits serious attention.

To avoid plagiarism, follow these simple guidelines:

1. *When in doubt, cite.*

   You may wonder how often you should cite a particular source. For instance, if you're using the same source for an entire paragraph, do you need to include a citation after each sentence? Here's a rule of thumb: if a reasonable reader could understand that two or three related sentences are from the same source, then citing at the end of the group of sentences is fine. However, citing just once at the end of a paragraph can lead to confusion regarding how much of the content is from the sources, and how much was written by you. Our advice is to cite enough so as to avoid any hint of ambiguity.

2. *Avoid quoting a source word for word.*

   In the humanities, it is common to include passages taken verbatim from outside sources such as novels or poems; these verbatim quotations must be included in quotation marks or somehow set apart from the main text. However, in science and engineering, verbatim quoting is much less common. The more usual route is to paraphrase your source using your own words (and still cite, of course). Note that using the exact words from a source without quotation marks, even though you include a citation, is still considered plagiarism.

3. *Keep track of your sources meticulously.*

   Programs like EndNote can be used to cite as you write, using whatever style your instructor specifies. If you write first without citing, and then go back and add citations, you may miss one or two citations and could end up with an unintentional case of plagiarism.

Summary table

Table 6.2 summarizes the guidelines on citing and listing references in this section and may be used as a guide to grading citations and references in undergraduate lab reports.

Table 6.2. Guide to citing and listing references in experimental engineering reports

Recognized citation style is used, usually the author-date style.	One citation style is used throughout except for unusual circumstance of occasionally citing numerous works.
All sources cited are listed in the reference section.	All references listed in the reference section are cited in the text.
Reference listing has complete bibliographical information.	Consistent capitalization system used in all listings of references.
Listing is in proper order with proper punctuation.	Italics are used for books and journal titles and quotes used around titles of reports and other shorter works.

CHAPTER 2.7

USING THE ENGINEERING LIBRARY: HOW TO CONDUCT RESEARCH IN THE ENGINEERING DISCIPLINES

Introduction

Up to this point it is entirely possible that the only search engine you have used to find information for your daily life as well as for your studies has been Google. And up to this point, it is entirely possible that this general search engine has met your needs.

However, as you advance through your engineering studies, Google will no longer suffice. You have reached a stage in your education at which you are expected to find and report on substantive, scholarly research conducted by professionals in a given field. The sources for this type of scholarly information are peer-reviewed journals, books published by academic and university presses, and formal reports issued by government agencies. But these are *not* the type of sources that are accessed in a basic Google search. Therefore, search engines like Google, Yahoo, and Duck Duck Go are, from this point forward, inadequate for the tasks you will be expected to accomplish.

So, if you can't "just Google it," how do you conduct research at this new level? The answer: using electronic engineering databases.

What is a Database?

A database is a collection of information (data) stored electronically. There are many different databases that collect information on wide range of topics, but what we are interested in here are engineering research databases. Engineering research databases are collections of citations of scholarly works (journal articles, books, technical reports, conference proceedings, for example) about all areas of engineering.

There are not only databases about engineering in general, but also specialized databases for the different disciplines of engineering like civil engineering, or mechanical engineering, or aerospace engineering. Moreover, there are even super-specialized databases. For example, within civil engineering alone, you will find databases that address specific areas of civil engineering like transportation engineering, geotechnical engineering, or structural engineering. Example 1 illustrates the different levels of specificity—from most general to most specific--possible for searching for information on transportation engineering.

Example 1:

General Engineering Database: Compendex

Catalogs resources for *all engineering disciplines*

OR

Specialized Engineering Database: ASCE Library

Catalogs resources related to *civil engineering only*

OR

Super-Specialized Engineering Database: TRID (Transportation Research Institute Database)

Catalogs resources related to *transportation engineering only*

Your job when doing research is to find the databases that collect the information most relevant to your research so you can efficiently access the scholarly material that will be most useful to you. The more specific the database you choose, the narrower the search results will be, and this is positive because it means the more efficient your search will be. To demonstrate what we mean by "efficient," look at Example 2. You will you see how much more useful the results get as the databases become more specialized.

Example 2:

Let's say we are studying transportation engineering and want to learn about roundabouts (commonly known as "traffic circles"), which are roadway alternatives to intersections. Table 7.1 shows the results when we used the words "roundabout design" in the three different databases mentioned in example 1. Results are reported for the first five entries in each of the successively specialized databases.

Table 7.1. Search results for "roundabout design."

DATABASE	SEARCH RESULTS
General Engineering Database: **Compendex**	*The first three entries had* <u>*nothing*</u> *to do with roundabout design. They provided citations on material fatigue, deviations from standard processes, and microwave dryers! Entries 4 and 5 did, indeed, relate to road roundabouts.*

DATABASE	SEARCH RESULTS
Specialized Civil Engineering Database: **ASCE**	*Four out of the five first entries were on road roundabouts. However, one entry was on the vibrations of the cables on a "Roundabout Flyover," which refers to structural engineering, not transportation engineering.*
Super-specialized Transportation Engineering Database: **Transportation Research Institute Database (TRID)**	*All five entries (in fact, the first 10 entries) were directly and specifically about roundabout design.*

This is not to imply that the general engineering databases are bad—far from it! In fact, if we just modify the general search words "roundabout design" to the more specific "roadway roundabout design" the *Compendex results give us five relevant citations*. The point is to demonstrate that if more specialized databases exist, they will be more efficient for your search because your search terms don't have to be highly specific to get meaningful results.

So, how do you find the databases that are appropriate for your research needs when in engineering alone there are several hundred databases that range from the general to the highly specific? The next section answers that question.

Working with Databases: How Do I Start?

Library Website:

The most important resource for working with databases is your college or university's library website. Notice that we didn't say library. This might come as something of a relief to you since there's a good chance you haven't even entered your college library. Or, if you have, there's a good chance you didn't go there to borrow a book but rather to work on projects with your fellow students. Twenty-first century libraries have changed dramatically from 20th-century ones, which were places students went to consult card catalogues (or later simple electronic catalogues) in order to find and check out the resources they needed.

Now libraries are becoming spaces *not* where students go to do research—they can do that from their dorm rooms—but where students go to meet to collaborate and use new, multimedia rooms. In fact, some libraries are moving books to offsite locations to free up space for collaborative work. Such a move acknowledges that the way people learn and the way they conduct research has fundamentally changed. The future (and present) is digital.

So, what do these types of changes mean for you? They mean that you will have to become proficient in not only conducting research online—which means learning how to find sources electronically—but also evaluating them for their suitability for your project before you even begin to write.

So, why did we say the most important resource is the university library website? Because the electronic databases that you need to conduct your research are accessed from the library website, not from a general search. This raises two important issues: First, these databases have been curated for academic work based on the expertise of librarians and input from researchers at the university. Therefore, they are your best bet for finding good resources. Second, these databases are subscription-based, which means they are not free to access. Luckily, your library holds the subscriptions on behalf of the students and faculty, and the links to these databases are housed on the library's website. As a student at the college or university, you have free access to these databases when you access them through your library's website. If you tried to access them directly (without going through the link on your library's webpage), you would be asked to login—and you don't have a personal subscription. They cost thousands of dollars.

Librarians

Even though we will be concentrating on a wide range of electronic databases that you can access from your own dorm room, don't forget about librarians! Remember, even though books may have been moved off-site, the librarians are still on site. Librarians are trained in the art of research and their help will always improve your research efforts. Many universities offer both in-person and on-line help. On-line help can come in the form of emails or real-time chat sessions. That being said, keep a reasonable timeline in mind when consulting a librarian. Librarians do not work on a fast-food model. Research takes time and thought, and therefore you can't just show up—in person or virtually—and expect to get the information you need in a matter of minutes.

But there isn't only one type of librarian; there are many types of librarians and people who work in libraries as part of the research enterprise. Therefore, if you seek the help of librarians, it would be useful to understand the range of assistance possibly available at your college or university. What follows is a list of the types of librarians possibly available at your home institution:

- General librarian
- Subject specialist librarian
- Media specialist
- Data visualization specialist

Additionally, it is possible that these librarians post electronic "Research Guides" on the library website that might help you in your work. What follows is a list of possible guides:

- Lists of the most relevant databases for conducting research in your field.
- Research guides designed for specific courses. In this case, the librarian will have worked with the course instructor to develop materials that will aid students in research projects for that *specific course*.
- Lists of the most relevant databases for conducting research in specific sub-genres of engineering.

And Finally...The Campus Library—or Libraries

Up to this point we have been talking about the campus library as if there were always only one library. Remember, many universities have several libraries on their campus—usually a main, general library and often several smaller specialized libraries devoted to specialized subjects like religion or engineering. When you begin your research, make sure you are using the library most appropriate to your studies.

As we mentioned at the beginning of this chapter, many campus libraries have become places for students to collaborate more than places to engage in research. That doesn't mean the physical library doesn't have much to offer the student-researcher apart from being the place where you can find a librarian and work with her or him in person. Far from it.

First, if your library still has books and journals on site—and most do—you may very well need these physical documents. For example, if a journal has not digitized its complete archives, and in your research you identify an "old" (pre-1980's) article that you need, you will have to go the library to access the physical copy of that journal. Whether your library has "open" stacks (you are able to access the book shelves directly) or "closed" stacks (you need a member of the library staff to retrieve your request), it (and not the internet) may still be the place where you find your resources.

Second, libraries often lend equipment like laptops and cameras, and you may need this equipment for your research project.

Third, perhaps your library has interlibrary loan agreements with other university libraries or even shared collections housed in an off-site repository (as mentioned earlier). In such cases, it is possible that you would visit your campus library to pick up the research material you ordered using the electronic databases.

Fourth, many libraries have exhibition space in which they showcase highlights from their collection or display artwork, and these displays can enhance the library experience.

What are the Best Engineering Research Databases to Use?

What follows is a list of general engineering databases. After that, lists of databases for the different disciplines and, in some cases, sub-disciplines, of engineering are provided. These lists will not necessarily reflect all the databases that your college or university subscribes to.

They are meant to serve as a starting point for your research and will give you an idea of the kind and range of resources that exist electronically. Some of these databases provide an abstract and index only and you have to request the document; some give you direct access to the document.

General Engineering Databases

1. Engineering Village: This is actually a platform that contains a range of databases within it
2. Web of Science: This is another platform that contains a range of databases within it
3. Compendex (EiVillage 2):
4. Scopus
5. IEEE Explore

Civil Engineering
Civil Engineering Database (ASCE)
TRID
GeoRef
GreenFILE

Mechanical Engineering
ASME Library
Inspec
Knovel
ASM Handbook

Materials Science and Engineering
Materials Science and Engineering Database

Aerospace Engineering
Aerospace Database
AIAA

Biomedical Engineering
BIOSIS
MEDLINE
PubMed

Systems Engineering
INFORMS Pubs Online
Computer Source

Electrical Engineering
Inspec
NTIS

Chemical Engineering
SciFinder
Knovel
Springer Materials

Example Assignment

It is not just laboratory reports or design reports that will require you to do research. What is reprinted below is a homework assignment that is designed to teach (among other skills) the fundamentals of report writing. What is interesting about this assignment in relation to our discussion about databases, is that this assignment would require students to consult many different types of databases in order to successfully complete the assignment. It basically answers the question, "Why would you ever consult a general database when you can consult a specialized one?" In this assignment, students would need to consult a specialized engineering database about transportation to learn about high-speed rail, but they would also need to consult databases on politics and social science to learn about community responses. Finally, they would need to consult the databases for contemporary news outlets to make sure they could accurately report on the current status of the project.

Short Report on the California High-Speed Rail Project

The California high-speed rail is a proposed north-south rail system designed to link major cities in CA (via the Central Valley) by means of high-speed rail service. The proposed rail project is highly controversial and will impact local and regional economies, communities, and environments in a variety of positive and negative ways.

You are to write a short (3-page) report on the proposed construction of the California high-speed rail project.

The objectives of this assignment are the following:
- To familiarize you with standard technical report format
- To have you practice synthesizing information from multiple sources
- To have you practice documenting electronic sources in a written document
- To introduce you to the concept of being responsive/non-responsive

Report Instructions:
1. Title the document.
2. Do not write more than three pages, double-spaced and 11pt font, excluding the References page. (You must call it "References.")

3. Use the following major headings in the following order:
   • Background: The CA High-Speed Rail Project
   • Proposed Functional Characteristics
   • Arguments in Favor of the CA High-Speed Rail
   • Arguments Against the CA High-Speed Rail
   • Status of the CA High-Speed Rail Project
4. Although you must use my exact headings—you will be penalized if you don't--you can create your own sub-headings (as appropriate) under any of the major headings.
5. Design the headings/sub-headings so that they are visually distinguishable from each other and the text.
6. Follow standard ASCE documentation style, including a list of References.

Research in Your Post-College Years

Your access to the wide range of engineering databases is one of the many advantages of being a student. But what happens to this advantage once you graduate? How can you access all of these databases once you are working? The short answer is, possibly you can't. As we mentioned earlier, these databases are subscription only, and these subscriptions cost thousands of dollars. Most engineering firms don't have the budget to support subscribing to all these expensive databases. But practicing engineers still must do research, and here are three ways they access the engineering literature:

1. Engineering firms often maintain subscriptions to specific engineering journals, not entire engineering databases. As an employee of that company, you can access the literature from those specific journals.
2. Government resources are publically available for all people, not just engineers. The Department of Transportation, for example, posts information on its website related to engineering research. They even have a specific section called "Tools for Researchers."
3. When you join (pay dues to) an engineering professional society, you will have access to any resources that it has purchased for its members.

Chapter 2.8

Preparing a Cover Sheet

Any long report and any report submitted to a library or document depository should include a cover sheet or title page. This page should include all the relevant bibliographical information so that a technical librarian can catalog and deposit the report without being forced to peruse the entire document.

According to the ISO [1], International Organization for Standardization, the title page should include the following information in the following order:

1. Any distribution limitation or security classification as described below.
2. Any report identifier such as a serial number issued by a particular organization or an ISBN number for a document published as a book.
3. The name and address of the responsible organization, which is essentially the publisher of the document.
4. A descriptive title, which should be concise but definitive. This should include any identification of a specific version (e.g., final or interim) and any revision or edition number.
5. The names of personal authors, using the full name of the author in natural order. The ISO recommends that some emphasis be added to identify the family name or other name by which the author should be addressed professionally. For example Drs. Burdell and Perez might use:

    George P. Burdell
    Jorge Perez de Burdelo

6. Omit any author block when the corporate author is the publishing organization.
7. The date of publication—meaning when the report is offered to the public or its intended recipient. To avoid any ambiguity, use the European convention on order and spell out or abbreviate the name of the month, as in 29 July 2015 or July 2015.
8. Date of priority in parentheses when appropriate as in articles submitted to technical journals. This is the date when the document or manuscript became essentially complete. It is included to help assert a possible right to the priority of discovery.
9. Any special notes such as approvals, disclaimers, and releases.
10. Repeat of distribution limitation.

Distribution limitation

A report may have a distribution limitation to restrict circulation or duplication to protect commercial trade secrets, patent disclosures, or other intellectual property rights. An example is a designation of "Proprietary Information" with accompanying limitations on use. Use this designation to inhibit someone who receives the report in the course of business from using or handling the information improperly, such as by adopting the information for his own use or by passing the information on to your competitors. Another typical limitation may be for the material to be maintained confidential and to be used only for a particular purpose such as evaluating a proposal. When national security issues are involved, a legally enforceable security designation may be imposed, such as "confidential," "secret," or "top secret." Proper handling of such a document is strictly prescribed by law and official regulations.

An example cover page that should be appropriate in an undergraduate course follows.

Reference

[1] ISO, "Documentation—Presentation of Scientific and Technical Reports," ed. Geneva, Switzerland: International Organization for Standardization, 1985.

PROPRIETARY INFORMATION:
FOR INSTRUCTIONAL USE ONLY
NOT FOR DISSEMINATION OR DUPLICATION

REPORT No. ME4055-2002-A4

Energy Release Evaluation of Cold Fusion Using Adiabatic Bomb Calorimeter
Final Report

Generic Institute of Technology
School of Mechanical Engineering
Experimental Engineering Laboratory Course
Summer Quarter 2002

Group A-4:

George P. Burdell
Georgia Prandtl-Burdel
Georges-Pierre Bourrel
Jorge Perez de Burdelo
Bu Tien

27 July 1999

(Manuscript Completed 21 July 1999)

Instructor Acceptance: _____ Date: _____ Final Grade: _____

PROPRIETARY INFORMATION:
FOR INSTRUCTIONAL USE ONLY
NOT FOR DISSEMINATION OR DUPLICATION

CHAPTER 2.9

GUIDELINES FOR GRAPHS

Introduction

Graphs are usually the most important exhibits in undergraduate experimental engineering reports. Graphs present your data in comprehensive and condensed visual format, and they reveal tendencies in the data and relations between independent and dependent variables. Since graphs are especially important, their preparation requires special care. Graphs prepared in spreadsheet programs, such as Excel, are common in student projects because spreadsheet programs are convenient and widely used. Plots prepared using other tools, such as Matlab or KaleidaGraph, are also acceptable for most instructors.

No matter what program is used to prepare a display, a graph must be legible and uncluttered, and its content must be clearly presented. Some especially important general tasks in designing and presenting a graph are these:

Specific Guidelines and Techniques

Graph type

In student reports for experimental engineering, graphs are usually expected to be so-called Scatter or XY graphs rather than line charts. The line chart is often used in business-oriented presentations that require all the points in a series to be evenly spaced along the horizontal axis without regard for the x value. In XY graphs, the horizontal position can be proportional to the x value; this is the usual convention for a mathematical graph, and it may better reflect the complexity of technical measurement and analysis.

Graph caption

In a report, a graph is presented as a Figure. Always caption your figures in reports with a unique consecutive figure number and a descriptive title. This caption is always placed below the figure. Be sure to cite every figure in the text. This rule applies to every exhibit or attachment. When an exhibit is cited by its specific number, its identification is usually considered to have become a proper name, such as "Figure 1.," and as such the name should be capitalized. Along with the unique identifier, use a descriptive title, such as "Figure 1. Velocity and Turbulence from Thermal Anemometer Traverse." Avoid a generic name or a mere restatement of the titles of the axes, such as "Figure 1. Velocity vs. Radius." You may be asked to print the caption above the graph when it is embedded in a spreadsheet page or when it will be displayed on a slide in an oral and visual presentation; however, most supervisors and most print editors require figures in written reports to have captions below them.

Lines and Markers

Usually, experimental measurements in mechanical engineering are best represented as a set of isolated data markers. Other data, such as the literature model in Figure 9.1 or a regression model developed from experimental data, are best represented as a continuous curve. Sometimes, literature data is presented in a table, such as a table of thermodynamic properties, but the table almost always represents merely a set of values computed at arbitrary points from a model, not a set of measured experimental data. Remember these differences and respect the ultimate product when selecting lines or markers.

SAMPLE FIGURE

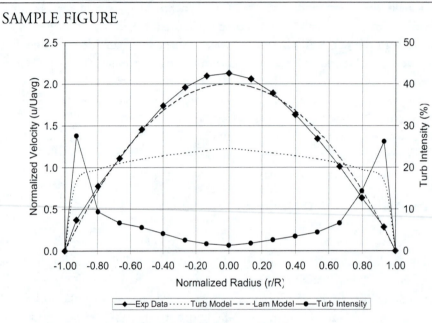

Figure 9.1. Example Plot with Two Data Sets and Two Models.

As shown in Figure 9.1, it is best to reserve markers for experimental data, such as the velocity data in that figure. Do not use markers merely to identify a line. Instead, use a distinct line type to identify a continuous curve, such as the model profiles from the literature in Figure 9.1 or for the regression model based on the data in Figure 9.2. The models may be obtained from an equation or they can be based on a table of values. Even if the model is based on a table of calculated values, use a continuous line to illustrate it. Otherwise, these markers can be confused with actual data points. In the unlikely event that a table of experimental data points from the literature is plotted, use markers for these data. On occasion, the experimental data will be a continuous, or nearly continuous, set of points, such as the time trace of an accelerometer. Such an experimental curve can be presented without markers since every point in the curve should be understood as representing an experimental datum. Otherwise reserve markers for data. Depending on the context, it is sometimes appropriate—and sometimes not appropriate—to connect the markers by straight lines as discussed in the next paragraph.

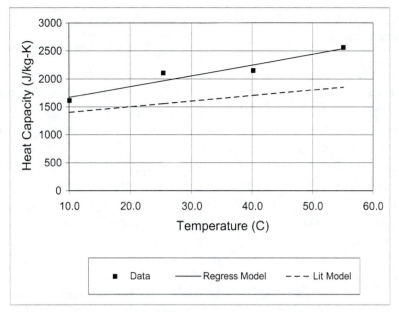

Figure 9.2. Heat Capacity Data (markers) and Regression Model (solid line) for Lubrication Oil Compared with Model from Tempco (1992, broken line)

Profiles or scatter plots

As illustrated in Figure 9.1, sometimes experimental data points should be connected by straight lines in a profile, particularly when multiple data sets are displayed on a single graph. As shown in the example, such connecting lines are mere tie lines and should be straight, not smoothed. In cases where a single data set is plotted, it is logical to display the data points as a scatter plot of isolated markers, illustrated in Figure 9.2. Specifically, data used to develop a regression model should be plotted as a scatter plot of unconnected markers with the regression model displayed as a solid line without markers. When feasible, an error band or uncertainty envelope for the regression model can usefully be plotted along with the regression line itself. In such a case, you should use one distinct line type, such as the dashed line, to indicate the upper and lower limits of the error band, as in Figure 9.3, which is presented later as a comparative example. As mentioned earlier, a comparison curve representing either a theoretical model or a regression result from the literature should not have markers; an alternative and distinctive line type is preferred.

Smooth lines or straight lines

The knowledgeable reader knows that no theoretical model or even a reasonable statistical model will pass exactly through experimental data. Consequently, lines used to merely connect markers into a profile, such as for the velocity and turbulence intensity in Figure 9.1, should be straight lines. This choice emphasizes that the lines are merely a convenient graphical way of grouping the data markers into a distinct set. A theoretical or statistical model is essentially a continuous set of points, and the appearance of the line drawn to represent this set may be improved by using the "smoothed line" property in Excel.

Identifying series in a graph

At least three methods are used to identify series. The simplest and most direct way is to identify the series in a legend. When you plot multiple data sets, it is wise to avoid using markers alone to help identify a line in your legend. However, markers can display indistinctly, so it is a good practice to augment the legend's data markers with representations of the line types that are used to group the data sets. An alternative or complementary way to identify different series is to place labels on the graph plane, the way the models are identified in Figure 9.1. There, the labels identifying the comparative models were added by inserting text boxes with the document edtior. A final method that may be suitable for a very simple graph is to identify the series in the caption. If only two or three series are plotted, notes in the caption (*e.g.*, parenthetical notes) as with Figure 9.2 can be used. Here, the figure caption is prepared in the document editor rather than the graphing program. The caption was placed below the graph as required for technical reports.

Use of colors

You should not rely on color to group your data sets. Instead, you should augment any color in your plots by using distinct line types or markers with distinct shapes to identify your different data series. You should do this because, while color printers are common, they are not universal, and many readers will view your work in black-and-white printouts. While color displays are acceptable and useful for screen displays, you should design displays that will be robust to black-and-white printing on low-quality devices.

Axes

Label and select the range for the vertical y and horizontal x axes carefully. A wide range may compress the data excessively. A compressed range may obscure an overall pattern by exaggerating minimal erratic variations. You should be particularly cautions about offsets that would omit or displace the zero point on either or both axes. Such an omission can distort proportions and obscure scaling relations such as those found in the pump and fan laws. Consider, understand, and respond to the significance of the (0, 0) point. If it is physically significant, such as with many normalized variables, manually adjust the axes to specify zero at the origin of an axis.

Secondary vertical axis

Graphing programs allow you to use a secondary axis to display a series with distinct units or when the magnitude of one series differs markedly from the others, as is the case for the data display in Figure 9.1; however, do not allow two confusing sets of gridlines to appear on such plots. In the unlikely event that it is appropriate to use two sets of gridlines, you should adjust the specifications on your graph to make the grids collinear. When you normalize your data correctly, you may be able to display several series on one axis without regard to units so long as the magnitudes do not differ too much. When two y axes are used, you must give each axis a distinct label and include the units for dimensional data.

Graph box

Whether to surround the graph with a text box as in Figure 9.2 or to eliminate this feature is a matter of personal preference if no specific guidance has been given. Avoiding the box eliminates one more feature to worry about and makes it easier to resize graphs without creating ragged right and left margins. Be consistent at least within each report on whether to include or eliminate the graph box.

Inserting separately printed graphs

On occasion, graphs may be generated by stand-alone instruments or data acquisition systems, such as oscilloscopes or anemometers, that may provide only screen-shot files or even printed hard copies. The very best practice in such cases is to generate a high resolution image with a scanner and insert the image as a picture into the report. Students should avoid using cell phone photographs; although cell phone cameras are often quite good, the light conditions in labs and computer clusters are usually poor, rendering student photographs unacceptable for reports.

Comparative examples

Figures 9.3 and 9.4 present the same data, and they offer an instructive contrast in terms of clarity and effectiveness. Figure 9.3 is clear and sharp, while Figure 9.4 has several basic deficiencies. The markers used to identify the literature model are misleading. They could be interpreted as data points that surprisingly, exactly fit the model curve from the literature. The x axis of Figure 9.4 is faulty. Since zero Celsius has no fundamental significance in this data set, there is no purpose in starting the horizontal scale there; the axis space to the left of 10°C is merely wasted. The extra title on the graph area in this figure is redundant. It is probably located correctly only for its intended use as a projection slide. For a graph in a report, the caption should be properly located below the figure. Either choice of vertical axis may be appropriate depending on the features of the data that are to be emphasized. Figure 9.3 emphasizes the difference between the data sets while Figure 9.4 tends to emphasize the reasonably good agreement.

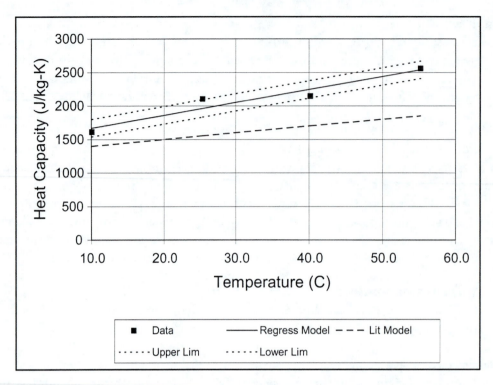

Figure 9.3. Heat Capacity Data (markers) and Regression Model (solid line) with Error Band (dotted lines) for Lubrication Oil Compared with Model from Tempco (1992, broken line)

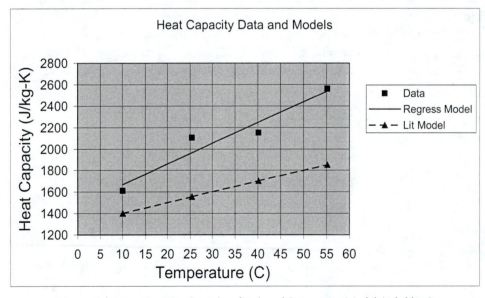

Figure 9.4. Heat Capacity Data (markers) and Regression Model (solid line) for Lubrication Oil Compared with Model from Tempco (1992, broken line)

The location of the legend block and a few other features in Figure 9.4 are also faulty. Since horizontal space is limited, the legend seriously crowds the graph. When it is feasible to do so, you should move the legend to an unused area of the plot plane or to the bottom of the display below the axis label. The gray tone printed on the graph area, which is the original default in Excel, is undesirable; it makes the graph hard to read in the original, much more so in a photocopy. It is always desirable to eliminate colors in the plot area in favor of a white background. The grid lines in Figure 9.4 are not desirable; you should always seek to eliminate, or at least to minimize grids in graphs. If grids cannot be entirely eliminated, it is best to minimize these, using only those lines that reveal something significant about your data.

Setting the default graph

In Excel, you can simplify the process of formatting plots by creating a default graph and saving it on your system. To define a suitable default chart, use some sample data and plot it using a reasonable chart layout from the Excel Chart Layout menu.

An XY chart is usually most appropriate. Set lines and markers to be black in color and distinct in type, and remove any default title lines. The axis titles should be set in 16-point type for most displays, because text is resized when graphs are inserted into documents. When you have finished formatting your plot, it might look like Figure 9.5.

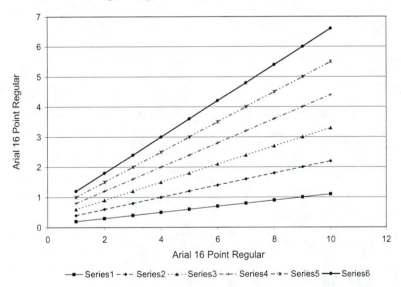

Figure 9.5. Representative Dummy Graph Created to be the Default Graph Type

Next carefully inspect your template graph, and examine the results after you insert it into a document and then print that document. Once you are satisfied with the dummy graph, you can specify it to be the prototype of a user-defined chart type by selecting the *Save as Template* option in your Chart Tools menu. Then the new user-defined type can be specified to be the default type in the program's *Insert Chart* dialog. With this user-defined chart, you can greatly reduce the time required for preparing graphs for your reports.

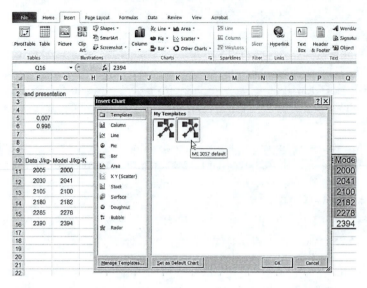

Figure 9.6. Dialog Box for Chart Type After the Typical XY Type Has Been Defined

Summary Table

Table 9.1 summarizes the guidelines in this section and can be used as a guide to grading graphs in under-graduate lab reports.

Table 9.1. Guide to preparing graphs for experimental engineering reports

Design is neat, uncluttered, legible, and effective.	Unique number is assigned and descriptive caption is given.
Always place caption below graph in written report.	Consistent capitalization system used in all captions.
XY graph, not Line chart, is almost always used.	Graph box included only with discretion and then consistently used in one report.
No reliance on color or subtle gray scales.	No shading of graph area.
Graph is cited in text.	Any cited graph is included.
Graph is inserted into text at first feasible location when possible or on separate page when necessary.	Graph is attached (*i.e.*, as one item appendix) to extended abstract only.
Full page graph on separate page in landscape orientation has caption to right. pasted on to graph page.	Separately printed graph is pasted onto page or has machine printed caption.
Zero point is purposefully and appropriately included or excluded in axes.	Secondary axis is included when necessary.
Axes are not so compressed or span is not so great that variation in data is not apparent.	Special care taken not to obscure scaling relations between normalized variables.
Axes are titled, with dimensions if existing.	Markers are strictly reserved for discrete data.
Data is presented in scatter plot or connected profile of markers as appropriate.	Distinct line types (*i.e.*, not markers) are used to identify continous data or model curves.
The series in a single series graph is identified only in caption.	All series are identified in legend, in notes or graph, or in captions.

CHAPTER 2.10

GUIDELINES FOR ILLUSTRATIONS

Engineering reports frequently include illustrations such as drawings and pictures. A complex drawing or photograph may be difficult to interpret. Complex drawings include engineering assembly drawings and artistic line drawings.

Simplicity is the most desirable feature in any figure. Any simple drawing, especially a schematic, such as Figure 10.1, will be better understood than a photograph, such as Figure 10.2, or a more complex drawing.

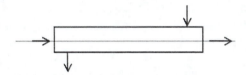

Figure 10.1. Schematic of Crossflow Exchanger Element

Even the rather complex schematic in Figure 10.3 is much more comprehensible than the photograph of the apparatus in Figure 10.2. A schematic may be better than an engineering assembly drawing or even an artistic drawing. Artistic line drawings can be very effective but these require skill, time, and expense.

When you communicate with clients or supervisors, it is usually best to avoid using a so-called engineering "detail" drawing, such as Figure 10.4, in a report. A detail is the name for a shop drawing with dimensions and manufacturing notes. Construction or manufacturing drawings are usually presented separately as an attachment, if the drawing is a single item, or in an appendix or separate bundle if a group of drawings is presented. If you must use a detail drawing in a report, it is probably best to simplify the drawing by removing some of the dimensional measurements. Figure 10.5 is an example of a simplified version of the detail drawing in Figure 10.4.

You should be critical of any proposal to use photographs in your reports. Digital cameras can produce excellent images, but technical photographs often require special lighting in order to display boundaries between components of devices. Without a heavy investment in lighting equipment, you may find that your drawings capture details and distinctions better than your photographs. If you do choose to include photographs in your reports, it is wise to pair your photographs with labeled diagrams that represent the same system or section as your photograph.

See the summary table following the example figures for other recommended guidelines.

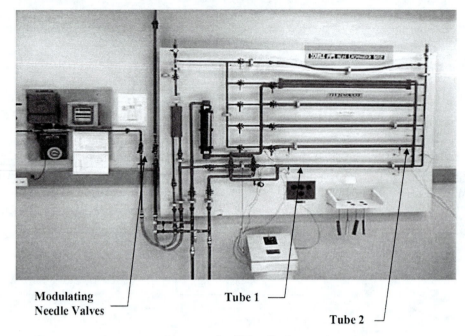

Figure 10.2. Photograph of Heat Exchanger Apparatus

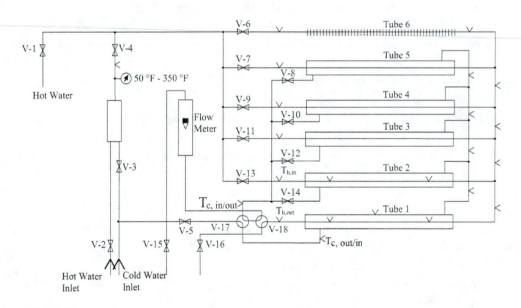

Figure 10.3. Schematic of Heat Exchanger Apparatus

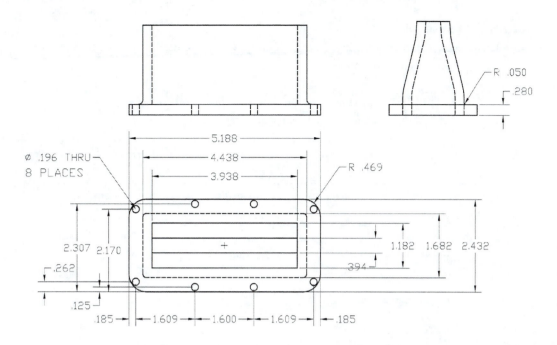

Figure 10.4. Engineering Detail Drawing of Experimental Nozzle

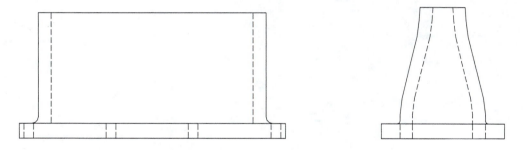

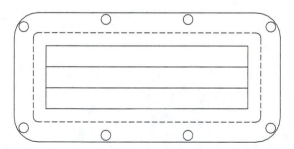

Figure 10.5. Simplified Drawing of Experimental Nozzle

Summary table

Table 10.1 summarizes the guidelines in this section and may be used as a guide to presenting illustrations in undergraduate lab reports.

Table 10.1. Guide to preparing illustrations or figures other than graphs
for experimental engineering reports

Figure is neat, uncluttered, legible, and effective.	Unique number is assigned and descriptive caption is given.
Caption always placed below figure or illustration in written report.	Consistent capitalization system used in all captions.
Photograph used only with discretion and usually accompanied by schematic.	Usually a schematic or simplified drawing is preferred.

Chapter 2.11

Guidelines for Equations

Introduction

Equations can be an important part of a technical report, as they are used to describe the analytical processes that underpin your judgments. In order to correctly use equations in your reports, you need, of course, to obtain and use equation editing software, and you need to understand how that software can best be used to capture your analytical steps.

Grammatical role

It is best to view an equation as a highly compressed sentence whose nouns have been compressed into symbols and whose verbs have been compressed into mathematical operators. You should write the text that accompanies an equation in such a way that the text and the equation work smoothly to describe your analysis step or steps.

Punctuation, numbering, and placement

While an equation represents a type of sentence, sentence punctuation is omitted to avoid introducing confusion into the symbolic display. Thus the display of Equation 11.1 is appropriate:

$$F = kx = m\,a \qquad (11.1)$$

where F is force, k is the spring constant, x represents deflection, m is the acceleration, and a represents acceleration

Note that the symbolic equation is centered on the line, and the equation number is placed flush with the right margin. The line immediately following the symbolic display defines each of the variables in the equation.

Only a few other simple comments are in order concerning the display of equations with the text of a report. Because equations present your efforts to meet an analytical goal, it is appropriate for you to describe that goal in a few words before you display each equation. Following the display and definition of variables, it is appropriate for you to state whether you have met your goal or whether further calculation steps are required. Based on these simple principles, we note that an equation is not expected to end a paragraph, and a series of equations should not be presented without intervening textual comments and explanation.

The details of equation editing are similarly simple, but a few guidelines that can save time and effort and minimize frustration are presented in the next section.

Entering and editing

Commercial text editors usually provide equation editing tools that produce acceptable results, and the free editor LaTeX provides first-rate equation editing tools. If your editor does not have a built-in equation editor, you will be able to find a variety of equation editors that are available as free downloads, or you may choose to use a commercial equation editor, such as MathType, that integrates well with a number of editing programs.

Most equation editors create a frame within your document, and embed the equation as a small graphical object. While highly developed, such editors do have a few confusing or awkward characteristics. These bugs can be problems to a novice, but as described in the next several paragraphs, they are easily managed once understood.

Math and text modes

In equations, as in text and tables, mathematical symbols should be displayed in Italic type. Some older equation editors assume that most symbols are mathematical unless they can be recognized as function names. The raw result of such a default is shown in the poor example of Equation 11.1 below; with the term "constant" selected and the non-math font selected, we obtain the improved example of Equation 11.2:

$$F > kx > ma, \; k > cons\tan t \tag{11.1}$$

$$Fsp = kx = m\,a, \; k = \text{constant} \tag{11.2}$$

For consistency and clarity, use italic type in the paragraph text when you refer to a math symbol such as *Fsp* for the spring force.

Parentheses

When you insert brackets or parentheses into your displays, it is best to use the parentheses template provided by your equation editor rather than to type the parentheses characters from the keyboard. This is because the equation display components automatically resize as you build up your equations, while characters typed from the keyboard cannot do this.

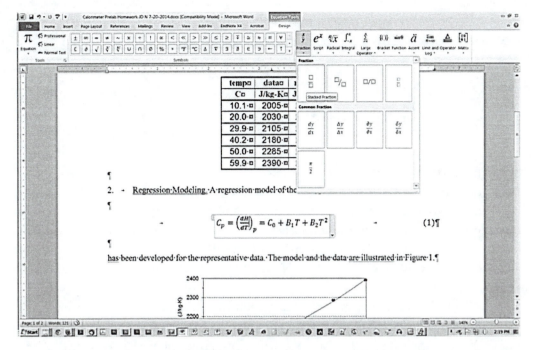

Figure 11.1. Document Window Showing Open Equation Window and Menu, with
Fraction Pull-Down Menu Open

Sizing or resizing

Equation editors are usually context-sensitive, which means that the characters in your equations will be set using the same font size as the surrounding text in your report. The few editors that do not do this automatically make it easy for you to select characters inside the equation display and modify their size. All equation editors make it easy for you to select individual characters or expressions and modify their style, including boldface or italic styles.

Summary table

Table 11.1 summarizes the guidelines in this section.

Table 11.1. Guide to writing equations in experimental engineering reports

Equation has logical grammatical setting.	Equation is neat, uncluttered, legible, and correct.
Equation is generated with Equation Editor not a line of text.	Equation is centered on separate line with unique number at right margin.
Equation is proper size, especially for projection slide.	Equation box is sufficiently wide to avoid truncation.
Templates from menu are used for parentheses, not symbols typed from keyboard.	Math symbols are in italics with functions, numbers, and notes in regular type.

Chapter 2.12

Guidelines for Spreadsheets

Introduction

Systematic organization of a spreadsheet or multiple page workbook will help make your numerical and graphical work easier and more reliable, and your spreadsheet will be easier to read or review or modify. Entire spreadsheets or even large blocks are usually too expansive and informal to be included in formal professional reports. For formal reports, you should extract a concise block of the spreadsheet and incorporate the block into your report as a table.

Even though a spreadsheet may not make a suitable exhibit, it must be organized well enough to produce reliable results. Organization is especially important if the spreadsheet must be reused or shared or used by others. In addition, spreadsheets are commonly included in less formal reports and most undergraduate lab reports to enable instructors to monitor the data and calculations. When you attach a spreadsheet, be sure to cite it in the text of the report. It is usually called an attachment, which is a one item appendix. An entire brief spreadsheet, or a pertinent block, may also be printed out and then physically pasted into a research notebook to create a secure documentary record.

To create an effective spreadsheet, organize the presentation functionally and design and format the entire spreadsheet so that it can be edited and updated efficiently, and so that it can be printed or inserted neatly. The organization described below and illustrated in the accompanying example has been found to be useful for most experimental work.

Functional blocks

For most experimental work it is effective and useful to organize the spreadsheet into the four sections described below. This organization is recommended for all undergraduate lab assignments, and it will be required for some courses. It is illustrated in the example that accompanies this chapter.

Header area

This very short section should include the name of the experiment, the names of the experimentalists, the course number and section, the time and date on which the experiment was performed, and the file name. This section should present enough information for a user to identify the purpose of this file.

Summary section

This typically short section should include only the most important summary results such as numerical value for an overall flow or a critical performance variable. This block should be placed near the top of the spreadsheet, so it will not be lost in a large spreadsheet or workbook.

Preliminary section

This section of medium length should include a list of constants, geometric information, and conversion factors that are used in the calculation of intermediate and/or final results along with any non-recurring calculations used in the data processing or presentation. **Importantly**, all unique data, such as experimental parameters, atmospheric pressure, calibration constants, and so forth, should be placed here so that this data can be comprehensively updated if necessary.

Data and calculations

This section should include the experimental data and the repetitive calculations for data processing or presentation. Only routine data should be placed here. You should use cell references to when you use unique data placed in the Preliminary Section. You should not hide unique data here, as that will make the spreadsheet hard to correct or update. This section may include an embedded graph of the data that one may or may not choose to include in any attachment appended to a report. It is probably good practice to embed a graph of the data when a block of the spreadsheet is printed to be physically pasted into a research notebook.

Placement of sections

Obviously, the header should be at the top of the spreadsheet. The summary and preliminary section should probably come next, but the order of these two sections is immaterial. The data and calculations section should probably come last to make it easy to extend this block if necessary. Each of these sections should begin with a prominent identifying heading.

Detailed guidelines

Inserting labels

Rows and columns should be labeled fully to identify data or results. In addition you should list the units in parentheses below or beside the column/row headings. Do not mix unit labels with your data. It is better to restrict the units to the column and row labels.

Formatting numerical data

To facilitate editing and to produce a neat appearance, use a consistent format for every column of data or block of related data. As shown in Attachment 12.1, all of the temperature data in column A are printed with one decimal place. Similarly, all of the vapor pressure data in columns B through F are printed with no decimal places showing. It is not necessary that each column use the same number of decimal places, but it is appropriate that you be consistent with decimals in each column.

Displaying decimal places

In a table, it is important to avoid displaying trailing insignificant digits. A spreadsheet is less formal, and it may be desirable to display trailing digits to track calculations or maintain consistent formats. Even so, it is good practice not to display an egregiously long tail of insignificant trailing digits. You must use judgment to determine what number of trailing digits is appropriate for your calculated values, and you must determine separately what number of digits is appropriate for your final result.

Formatting for printing

Use good judgment in formatting and printing a spreadsheet or a portion of a spreadsheet. Only when required should you print the entire contents of a spreadsheet, and then you should append it to your report as a single item appendix; it can be called an attachment. Such extensive attachments usually are not acceptable in formal professional practice except for internal use. A similar use is in an undergraduate lab course where data and intermediate calculations must be displayed.

It is usually best to extract a small and specifically designed block of pertinent data from the spreadsheet and then refine it for presentation as a table. If a spreadsheet attachment is used, it should fit onto a single page or on a single page with continuing pages that have their own adequate organization.

Printing spreadsheet as separate item

The spreadsheet, or at least its important blocks, may be printed as a separate item and appended to the report as a separate attachment. When you do this, you should include the unique attachment number and descriptive title in a header printed above the spreadsheet block.

Examples

The simple examples, Attachments 12.1 and 12.2, show the unique identifying number and descriptive title, printed as a header. Also shown are the heading, the preliminary section including general data, the summary section, and the section of routine data and calculations in the spreadsheet itself. Note that the title is printed as a header so that it will appear centered and prominent. Use of the header is especially desirable in order to display the heading on every page when a multiple page printout is needed.

Summary

Spreadsheets should be organized into four functional and logical blocks to include the heading, the summary, the block of unique data and one-time calculations, and the block of routine data and repetitive calculations. A working spreadsheet is usually too extensive and too informal to be included in a widely published report. It is better to extract pertinent data into concise tables and integrate the tables into the text. In a less formal report, you may include a spreadsheet as an attachment that can be printed separately or preferably inserted as a picture attached to the main report. When a multiple page spreadsheet must be attached to a report, pay special attention to its organization for printing: print only the necessary blocks and avoid pages with dangling or poorly identified columns or rows. Pictures of spreadsheets or separately printed versions attached to informal reports, such as student reports, should include all important blocks and have a unique attachment number and descriptive title in the printed header. In all cases, the spreadsheet should be functionally organized, neatly formatted, and well identified. Table 12.1 summarizes the guidelines for spreadsheets as discussed in this chapter.

Table 12.1. Guide to preparing spreadsheets and incorporating them
into experimental engineering reports

Overall design and detailed format are neat, uncluttered, legible, and effective.	Unique number is assigned and descriptive title is given in heading.
Spreadsheet is presented as an attachment cited in text.	Concise block of pertinent data is also extracted and presented in a table.
Title heading is centered and prominent with type size at least 10 point.	Spreadsheet is structured with heading, preliminaries, summary, and data blocks.
Heading identifies purpose and history of spreadsheet.	Unique data, constants, and parameters are placed in the preliminary section.
Summary section is near top of spreadsheet.	Routine recurring data are in a well-organized separate section.
Consistent numerical format is used in a related block or blocks of data.	Units are identified for every dimensional data block, typically in column heading.
Recurring data are identified, usually with column header.	Unique data are identified, usually with identifying phrase.
Printed margins are adequate and at least 1 inch all around.	Multiple page printout is carefully designed especially in continuing pages.
Block of formulas presented in separate attachment if required.	Optional concise summary regression block prepared if appropriate.

1. Print the title as a header in the spreadsheet so that it will appear centered and prominent on every page, or insert the title as a heading in the document above the picture of the spreadsheet as in this example.

Attachment 12.1 Experimental Spreadsheet for Vapor Pressure Lab[1]

file: PVAPOR99   21 June 2013   S. M. Jeter
Spreadsheet for processing vapor pressure data for CFC-12.

Summary Results
          0.9943424 = R squared
          0.8501812 = alpha risk

Constants and Parameters
          Calibration data: 1.0 = scale
                                     0.0 = offset (kPa)
               97.9 = atmospheric press (kPa)

Experimental Data

Temp C	Pv Data kPa	Pv Model kPa	Pv Lit kPa	Pmes kPa	Pcorr kPa	1/T 1/K	1/T^2 1/K^2	ln (Pv) Data
30.1	748	756	746	650	650	0.003298	1.087E-05	6.617269
45.2	1098	1061	1062	1000	1000	0.003141	9.867E-06	7.001155
59.9	1398	1452	1519	1300	1300	0.003003	9.015E-06	7.242726
75.0	1998	1971	2082	1900	1900	0.002872	8.250E-06	7.599852

Concise Quadratic Regression Block

Constant		17.094807
Std Err of Y Est		0.053826
R Squared		0.994342
No. of Observations		4
Degrees of Freedom		1
X Coefficient(s)	-4229.962	320157.3
Std Err of Coef.	8243.672	1335225.3
t-stat		0.239778
alpha		0.850181

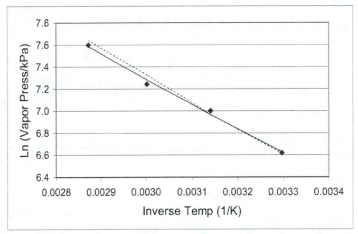

[1]Print the title as a header in the spreadsheet so that it will appear centered and prominent on every page, or insert the title as heading in the document above the picture of the spreadsheet in this example.

194

Attachment 12.2 Block of Experimental Spreadsheet Showing Formulas Used for Vapor Pressure Lab

file: PVAPOR99   21 June 2013   S. M. Jeter

Spreadsheet for processing vapor pressure data for CFC-12.

Summary Results

       =Sheet4!B5   = R squared

       =Sheet4!E19   = alpha risk

Constants and Parameters

	Calibration data:	1	= scale
		0	= offset (kPa)
	97.9	= atmospheric press (kPa)	

Experimental Data

Temp C	Pv Data kPa	Pv Model kPa	Pv Lit kPa	Pmes kPa	Pcorr kPa
30.1	=F20+$B$13	=EXP(J20)	745.8	650	=$D$11*E20+$D$12
45.2	=F21+$B$13	=EXP(J21)	1062	1000	=$D$11*E21+$D$12
59.9	=F22+$B$13	=EXP(J22)	1519	1300	=$D$11*E22+$D$12
75	=F23+$B$13	=EXP(J23)	2082	1900	=$D$11*E23+$D$12

GUIDELINES FOR TABLES

Introduction

Tables allow you to present extensive quantitative information in a limited space. To be effective, tables should be carefully designed, well integrated, and concisely formulated. Table 13.1 is a reasonable example. As dictated by most publishers' style manuals, the descriptive title is placed above a table, as shown in the accompanying example. The left hand column usually displays the independent variable (for experimental results) or a list of categories (for informational displays). The first row usually displays an array of text entries called "heads" that identify the columns and that include units when these are being used. You should use nouns or noun phrases as heads. In preparing heads, pay attention to formatting and alignment. Frequently the heads are centered horizontally and vertically, and the quantitative data are either aligned by their decimal points or, if whole numbers are used, justified to the right. Be sure the data are internally consistent, as a critical reader can easily check tabulated data. When your tables are very long but only a few cells wide, it may be acceptable to display them in multiple columns. If you do this, be sure that your heads are reproduced at the top of each new column.

Table 13.1. Estimate of bias or uncertainty B in shaft power measurement

Measurement	U_x[a]	Influence Coefficient, $\dfrac{\partial \dot{W}}{\partial x_i}$	$U_i^2 = \left(U_{xi} \dfrac{\partial h}{\partial x_i} \right)^2$	Basis	Source
shaft speed, N	10.0 RPM	$\dfrac{\dot{W}}{N} = \dfrac{7050\ \text{W}}{1020.\ \text{RPM}} = 6.91\ \dfrac{\text{W}}{\text{RPM}}$	4800 W^2	resolution	(1)
arm length, r	2.0 mm	$\dfrac{\dot{W}}{r} = \dfrac{7050\ \text{W}}{307\ \text{mm}} = 23.0\ \dfrac{\text{W}}{\text{mm}}$	2100 W^2	measurement	(2)
force, F	5.0 N	$\dfrac{\dot{W}}{F} = \dfrac{7050\ \text{W}}{215\ \text{N}} = 32.8\ \dfrac{\text{W}}{\text{N}}$	27000 W^2	calibration	(3)
		sum of $U_i^2 =$	33,900 W^2		
		Expanded Uncertainty B[b] $=$	180 W		

Sources: (1) physical inspection, (2) precise measurement, see text, (3) calibrated by manufacturer (1997)

[a] Expanded Uncertainty B in individual direct measurement, b 95% confidence limit

Design and editing

Numerical formats

Numerical data must be in a format consistent with the number of significant digits. Usually a consistent format (*i.e.*, fixed number of decimal places in fixed point or scientific notation) can be used within a column, but you may break the pattern if necessary to avoid displaying insignificant digits.

Numerical format for statistics

Statistics are indirect measurements or numbers calculated from more direct measurements. It is possible to evaluate the uncertainty in any indirect measurement from the uncertainty of the underlying more direct measurements, and these uncertainties should always be known. However, this evaluation can be complex and is usually avoided except for the parameters in regression models. However, this creates a problem in the way we track and display significant digits, as the actual number of significant digits in other familiar statistics, such as the R-Squared or the alpha risk in regression analysis, is usually not known. Numerical analysis shows that these statistics can have many significant digits even when the data have few significant digits. To solve this problem in class reports, it is reasonable to present statistics that have indefinite uncertainties, such as the R-Squared or the alpha risk, with enough digits to make any necessary comparisons. For example, you can reasonably use enough decimal places to distinguish two R-Squared values (*e.g.*, .9996 < .9998) even if the underlying direct measurement data have fewer significant digits.

Gridlines

For class reports, it is best to create tables that have clear gridlines to separate the columns and the rows.

Table footnotes

As illustrated in Table 13.1, tables may be immediately followed by footnotes. In particular, an overall source footnote to acknowledge the origin should follow any table that is entirely or mostly borrowed. Similarly, the source of data for any part of the table should be defined with a specific source footnote. Engineering tables frequently include very brief explanatory notes actually within the table. Longer explanations can be written in a footnote or can be included in the text with a reference in a footnote. Such source or explanatory foot-notes that refer to parts of the table may be indexed with a capital letter in parentheses, as illustrated in the example above. Footnotes explaining or identifying a specific entry may be indexed with a lower case letter in superscript, as also shown in the example. Letters are preferred to numbers as indices to table footnotes.

Cell format

Technical tables commonly begin as spreadsheet blocks that are then copied and pasted into the report document. As shown in the first column in example Table 13.2 below, the resulting format has the numbers flush to the right. This alignment is only marginally acceptable in a report. A somewhat better alignment is used in the second column where the data is centered in the text using the toolbar icon, although the uneven numbers of digits on the left created problems with alignment on the decimal point. A better alignment is used in the third column. Here a right indent of 0.1 inch has been set, leaving a little space to the right of the number. Either alignment in the second or third columns is acceptable but not optimal. The results in the fourth and fifth columns present a somewhat more professional appearance. However, in both columns, the number of decimal places in the last three rows have been adjusted to avoid displaying insignificant digits, so here, too, the decimal points are out of alignment. The last column is optimal. Here, the decimal tab is used to align the decimal points.

Table 13.2. Examples of various formats in table cells

angle(rad)	angle(rad)	angle(rad)	angle(rad)	angle(rad)	angle(rad)
3.1416	3.1416	3.1416	3.1416	3.1416	3.1416
6.2832	6.2832	6.2832	6.2832	6.2832	6.2832
12.5664	12.5664	12.5660	12.566	12.566	12.566
25.1328	25.1328	25.1330	25.133	25.133	25.133
50.2656	50.2656	50.2660	50.266	50.266	50.266
default from spreadsheet	centered	right indent paragraph format	centered	right indent paragraph format	decimal tab stop used

Summary table

Table 13.3 summarizes the guidelines in this section and can be used as a guide to editing and grading tables in undergraduate lab reports.

Table 13.3. Guide to generating tables for experimental engineering reports

Table is cited in text.	Table is concise, neat, uncluttered, legible, and internally consistent.
Table has unique number and descriptive title in heading.	Table is centered horizontally on the page.
No insignificant digits are displayed.	Data is identified usually in column headers with units as existing.
Table is limited to one page.	Numbers are properly aligned.
Consistent numerical display format used except as limited by significant digits.	Scientific notation used for very large or very small magnitudes.
Appropriate text format, usually centered, is used for headers and notes.	Source and descriptive notes and footnotes are provided.

CHAPTER **2.14**

GUIDELINES FOR LISTS

Technical writing can often be better organized and condensed by using lists. Two types of lists are common: "vertical" lists in column format, and lists in text style called "run-in" lists. Vertical lists are common in this manual. Both types help organize and simplify technical writing.

A run-in list is merely a series within a sentence or paragraph, with the items numbered in parentheses and separated by commas or periods with a final conjunction (*e.g.*, "and" or "or") as in ordinary writing. You should insert minimal or no additional punctuation in such a list.

Vertical lists are common in technical writing. A vertical list is in column format, and the elements may be indexed with numbers or marked with "bullets" such as "•." Usually you should omit terminal punctuation, whether commas or semicolons, in a vertical list, unless it is a list of sentences.

All lists of both types must be "rhetorically parallel," meaning that the elements should be logically and grammatically similar, such as all nouns in a list of the components of a system or all imperative sentences as the steps in a procedure. Rhetorical parallelism is a grammatical similarity used to emphasize conceptual, and usually physical, similarity.

A Bullets and Numbering utility is available with most commercial editing programs, and is especially helpful in composing a single column vertical list such as the following:

1. Preview lab manual and lecture notes.
2. Attend preparatory lecture and preview.
3. Complete any pre-lab assignment.
4. Arrive at lab promptly.
5. Conduct lab with attention to details.
6. Gather data, make notebook entries.
7. Process, analyze, and evaluate data.
8. Outline report in notebook.
9. Draft report, emphasize exhibits.
10. Proofread and edit report.
11. Use editing guidelines as final checklist.
12. Submit complete report on time.

Obviously this utility makes a single column list almost automatic. Refining the paragraph format to increase the right margin is the only slightly sophisticated step.

Table utilities are handy for creating multiple column lists. Multiple columns are helpful for lists of brief items to avoid a very long narrow list. When you make a list using the Table utility, you should omit any bordering lines in or around the cells of the table. The following example reproduces the above list using the Table method to make a multi-column vertical list.

1. Preview lab manual and lecture notes.
2. Attend preparatory lecture and preview.
3. Complete any pre-lab assignment.
4. Arrive at lab promptly.
5. Conduct lab with attention to details.
6. Gather data, make notebook entries.
7. Process, analyze, and evaluate data.
8. Outline report in notebook.
9. Draft report, emphasize exhibits.
10. Proofread and edit report.
11. Use editing guidelines as final checklist.
12. Submit complete report on time.

It is best not to end a paragraph or a section with the last item of a list. Such a termination is much too abrupt of a halt. Use a closing sentence at least, both as a transition and as a signal that the list is complete. In every case, be sure that a list is embedded in a paragraph with at least one preceding sentence to introduce and identify the list and at least one following sentence to complete the paragraph.

A list allows a complex paragraph or section, burdened by minutia, to be simplified and improved in organization by itemizing the details in a list. Good applications for lists, at least in informal reports, are steps in a procedure or components in an apparatus. Maintain rhetorical parallelism in all such lists. For examples, the steps can be imperative sentences, and the components can be noun phrases. Even in formal reports a list is often used in the closure to itemize the significant findings and conclusions. Obviously lists can be useful tools in efficient technical writing.

Summary table

Table 14.1 summarizes the guidelines in this section and will be used as a guide to grading lists in undergraduate lab reports.

Table 14.1. Guide to preparing lists used in experimental engineering reports

Lists used when appropriate and helpful to simplify and enhance presentation.	List not used for incommensurate items.
Run-in list has no extraneous punctuation.	Vertical list can have bullets, numbers, letters, or no indices as appropriate.
Use multiple column list for brief items.	Optional right indention of vertical list is helpful.
List must be rhetorically parallel.	Consistent capitalization system used in all lists, usually initial capital only.

Chapter 2.15

Integratng Graphics into Written Documents

Introduction

As you have seen in the examples of reports in previous chapters, engineering reports—no matter the type—contain a lot of graphical elements. Laboratory reports often include schematics or photographs of experimental set ups in the Methods section, and they certainly include graphical data in the Results section. Design reports typically include photographs or drawings of sites (civil engineering) or devices (mechanical engineering), and they certainly include graphical displays of the recommended design solution.

But how do you do incorporate the graphical elements correctly? Are there hard and fast rules about what graphics should look like, or are there just guidelines and suggestions? Are there standard practices for integrating graphics into a written document, or do you just put the graphics where they fit? And how about describing and interpreting the graphical elements? How do you write about graphics so that your readers can understand what they're looking at?

Knowing the answers to these questions is central to your ability to create a high-quality engineering report because data displays are central to these reports. They are, in fact, a defining feature. Therefore, you need to know how to work correctly with visual data. Your responsibility regarding visual data in engineering reports can be broken down into three tasks:

TASK 1: Collecting the data.

TASK 2: Displaying the data according to standard practices and standards of excellence.

TASK 3: Integrating the data into a written document and providing the associated verbal descriptions and analyses.

Task 1—collecting the data—is something you will learn in class. If you are in a laboratory class, the instructor or TAs will teach you how to set up the experiments, how to run the equipment, and how to record and analyze your findings. If you are in a capstone design class, the instructors, TAs, or your project's sponsors will teach you how to locate the information you need in order to create a viable solution to the engineering problem at hand.

Task 2—displaying the data—is discussed in several chapters in Part 2 of this book. Individual chapters offer guidance about designing graphs, tables, and schematics (illustrations). Please refer to the standards presented here when designing your data displays.

Task 3--integrating the visual data into a written document and providing the associated verbal descriptions and analyses-- is what this chapter addresses. However, before you begin this step in your report-writing process, make sure that you consult the various chapters on data design in Part 2. There is no value in incorporating visuals into a written document if those visuals do not meet basic, established standards of excellence.

Identifying Figures and Tables

Figures: There are many different types of figures: line graphs, bar graphs, schematics, maps, photographs, and pie charts, among others. However, when we caption them, we call them all generically "figures." You will *never* caption a visual display as "Graph 1," "Plot 1," or "Photograph 1." It will always be identified as "Figure 1." This format is considered a standard practice. That being said, you can certainly call a map a map in the text of the document, and if you write about the markers or symbols used to distinguish data sets in a graph, you can call it a graph. The rule about using the term "figure" applies exclusively to the caption.

Tables: The displays that have columns and rows are called tables. Unlike figures, there is not a wide variety of displays that are called tables. Tables are always displays with just columns and rows. What goes inside the cells of the table can be numbers or words.

Now that you understand the difference between figures and tables, here are some fundamentals for incorporating these displays into a written document.

5 Rules for Integrating Graphical Elements

What follows are five basic rules (and associated discussions) for integrating graphics into a written document.

Rule 1: All graphical elements must be introduced in the text BEFORE they appear. You can continue talking about these elements after they appear, but they have to be introduced first.

Discussion:

There are three ways you can introduce graphical elements in a text.
   1. Figure 1 shows... or Table 1 shows...
   EXAMPLE: "Figure 1 shows a schematic of the experimental set up."
   2. ..., as shown in Figure 1 or ...., as shown in Table 1.
   EXAMPLE: "The experimental set up has five components, as shown in Figure 1."
   3. (Figure 1 or see Figure 1)
   EXAMPLE: "The experimental set up has five components (see Figure 1).

The one way you DO NOT introduce a graphical element is by writing about "the following figure" or "the following table." *The graphic has to be specifically identified by its number.* This system makes it possible for readers to find the graphic referred to in the text if it does not happen to appear on the same page where it is introduced. Additionally, you *do not* include the phrase "below," as in the phrase "Figure 1 below." It is simply unnecessary.

Rule 2: All figures are numbered and captioned BELOW the figure, like this:

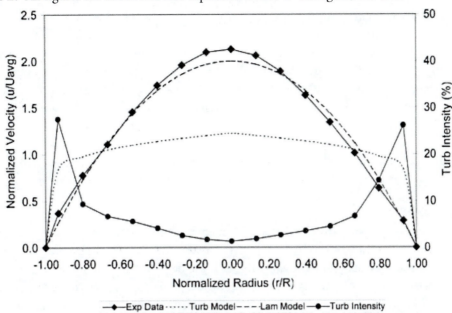

Figure 1. Relationship between Laminar Model and Turbulent Model.

Discussion:

This format is considered a standard practice, but you will inevitably see publications in which the figure number and caption are to the side of the graphic or even above the graphic. Publications have their own editorial policies, and they can do whatever they like. Moreover, if you ever send an electronic manuscript to a publisher, it is often the practice to send two separate files—the text file and the graphics file. Once these two elements are decoupled, you are no longer in charge of how they appear in the text. That being said, standard practice is to place the figure number and caption below the figure. Therefore, for documents you create as a student the standard practice of placing the figure number and caption below the figure is recommended.

Rule 3: All tables are numbered and titled ABOVE the figure, like this:

Table 1. Heat Leak Data and Calculated Conductance.

Air Speed (m/sec)	Electrical Power (W)	Air Temp (C)	Surface Temp (C)	Experimental UA (W/C)	Regression UA (W/C)	Error (%)
13.9	5.0	19.0	34.3	0.327	0.331	1.4
9.0	5.0	16.6	36.2	0.255	0.258	1.1
9.0	10.0	17.4	54.5	0.270	0.257	-4.6
4.5	5.0	18.1	45.0	0.186	0.191	2.7

Discussion:

This is considered a standard practice, but you will inevitably see publications in which the table number and title are to the side of the graphic or even below it. However, such unconventional layouts are not as common as they are with figures. But as noted in Rule 2, publications have their own editorial policies, and they can do whatever they like. That being said, standard practice is to place the table number and title above the table. Therefore, for documents you create as a student the standard practice of placing the table number and title above the table is recommended.

Rule 4: Integrate the graphical element as close to its introduction as possible, but make absolutely sure it is inserted in the same section in which it is introduced.

Discussion:

When integrating graphics into a written document, the goal is to create a document that makes it easy for the reader to go back and forth between the text and the graphic. The closer the graphic is to its introduction, the easier this task will be for the reader. Sometimes page layout or the size of the graphic makes it impossible for the graphic to appear on the same page as it is first mentioned. If this is the case in your document, just make sure to place the graphic as close as possible to the first mention and make sure the graphic is identified by a number and that that number matches the figure number you have assigned it. (Sometimes poor editing results in a mismatch between what you identify as Figure 1 and what is actually labeled Figure 1.) Whatever the layout, make absolutely sure that the graphic you mention in, say, section 2.1 of your report, appears in section 2.1, not in section 2.2.

Rule 5: Don't introduce another figure before you've displayed the first—*unless* they are related—as in the phrase "figures 4 and 5 show" or as demonstrated in the following example from Part 1 of this book:

EXAMPLE

"Whereas Figure 1.1 shows a sketch of a signal with zero DC offset, Figure 1.2 is a sketch of a signal with a 3-volt DC offset. Note that the vertical scales in the two figures are different.

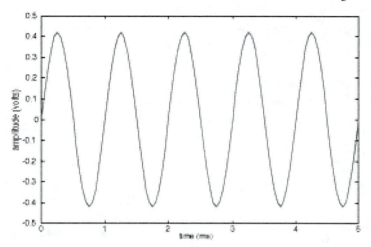

Figure 1.1. Sketch of a sinusoidal (AC) voltage with no constant (DC) offset.

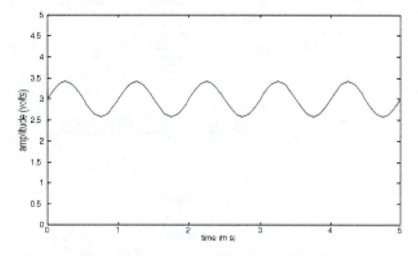

Figure 1.2. Sketch of same sinusoidal (AC) voltage signal
shown in Figure 1.1 after a 3-volt DC offset has been added.

How to Write the Verbal Descriptions and Analyses That Accompany Graphics

You have now learned how to introduce graphics and integrate them into a written document. This is good, but you're not done yet. You have to do more than just copy data from an excel spreadsheet, introduce it, and paste it into the appropriate section of the engineering report. Your job now is to provide meaningful and elegant verbal descriptions of the graphical information so that readers can a) actually see the data you are describing; and b) move their eyes easily back and forth between the text and the graphic and not lose their place as they return to either one.

What follows are two examples (one figure and one table) of data and their associated descriptions. A discussion of the written description follows each example.

Example 1: Figure (schematic):

Figure 1 shows a schematic of a recirculating system. The system is used to determine head loss characteristics of a test ball valve.

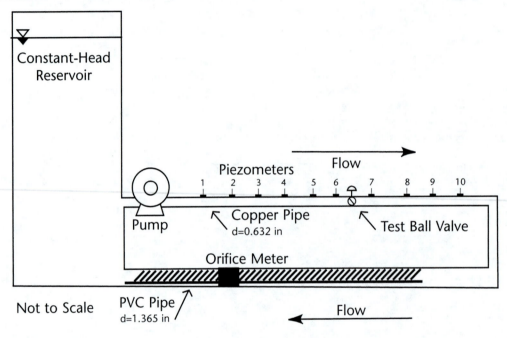

Figure 1. Schematic of a recirculating system.

The water circulation begins in the constant head reservoir. Water then travels through a pump that discharges the water at a constant rate into a copper pipe. Piezometers located downstream from the pump measure the piezometric head. The test ball valve is connected to the copper pipe between piezometers 6 and 7. The copper pipe is connected to a PVC pipe, which houses the orifice meter, and this PVC pipe section completes the circulation loop back to the reservoir.

Discussion:

This example shows the introductory text, the inserted graphic, and then some additional, explanatory text after the graphic. First note that the schematic reads left to right. That helps us in our effort to describe it, because we read the schematic in the same direction that we read text. Second, note that that the first sentence ("Figure 1 shows...") matches the figure caption. If the figure caption and the way you identify the figure in the text don't match, you can potentially confuse your reader. Third, notice that the clearly labeled parts of the schematic aid you as you go back and forth between the image and the text. That is one reason why a well-designed figure is the first step to a well-written verbal description. Finally, notice the ease with which you can go back and forth between the figure and the text: all components are mentioned according to their position in the system process.

Example 2: Table

Table 1 displays the speed data collected from in-vehicle GPS equipment.

Table 1. Speed Data Collected from In-Vehicle GPS Equipment

Date	Time (EST)	Latitude	Longitude	Speed (m/hr)
02.14.18	0900.00	40.7490	74.3880	0.00
02.14.18	0900.10	40.7491	74.3881	20.00
02.14.18	0900.20	40.7499	74.3890	30.00
02.14.18	0900.30	40.7515	74.3902	40.00
02.14.18	0900.40	40.7525	74.3925	50.00

Based on the information displayed in Table 1, the reader can see that all of the data were collected on February 14, 2018, over the course of 40 seconds: between 9 am and zero seconds and 9 am and 40 seconds. Looking at the final column (speed), the reader can see that although the vehicle was not moving at 0900, by 0900.40, it was traveling at 50mph.

Discussion:

This example also shows the introductory text, the inserted graphic, and then some additional, explanatory text after the graphic. Notice how the explanatory text directs the reader to read the graphic from left to right—the way English text is written. Also notice that even though the point of the table is about speed data (the table title indicates that), the text takes the reader methodically through the table—left to right—and doesn't make the reader's eyes dart all around looking for the information. Notice also, that the description left out two columns—latitude and longitude. These values, while important to the data-gathering enterprise and always recorded by this software, isn't important for the point of the display: speed data. Therefore, the writer was comfortable leaving it out of the description in order to streamline the text and make it easy for the reader to see the important data—the speed.

Chapter 2.16

Policies for Notebooks

Introduction

The research notebook is an important tool for organizing and preserving your data, ideas, and observations. The notebook exists as a tangible record of priority in patent proceedings or as documentary evidence of good practice in cases of research integrity or product or professional liability actions. It should be brought to, and used in, every lecture and lab session.

Notebook media requirements

In graduate research or professional practice, the research notebook should be permanently bound, with machine printed page numbers. For convenience and economy, students may be allowed to use an ordinary 8.5 × 11 inch composition notebook in the undergraduate laboratories, and they may be allowed to number the pages by hand. While this is inappropriate for a professional research notebook, it is acceptable for introductory undergraduate laboratory classes. You should clearly identify the notebook by writing your name and affiliation (*i.e.*, your particular course) on its cover. You should then consecutively number by hand all pages that you intend to use. Number the front and back of each page if you write on both sides of a sheet. You need not number all of the pages in your class notebook; you may wish to preserve unused pages for your next laboratory course.

Notebooks are used to establish integrity in records, so permanence is critical for all entries. Consequently, you should record all entries in **ink** rather than pencil. If you make an error, you should strike through it with a single line, as here: ~~errors.~~

You should avoid leaving large blank blocks or gaps in the notebook. To demonstrate that an area was blank when you finished using a notebook page, it is best to strike or scribble through any unused blocks. This practice is considered to prevent fraud because it prevents others from modifying your work after it has been completed.

If exhibits like graphs or printouts from instruments are inserted into the spreadsheet, they should be permanently fixed into the notebook as well. If you use tape instead of permanent adhesive to secure exhibits, initial over the edge to preclude tampering.

Special requirements for coursework

Lab faculty usually impose a few special requirements to facilitate instruction in the areas of lecture notes, demonstration notes, and outlines. You might be required or recommended to use the notebook for lecture notes that do not fit into space reserved in any pre-printed notes that you may have received. In particular, include any special instructions that may have been issued after your notes and handouts were prepared. Additionally, advance equipment demonstrations are sometimes conducted, and these demonstrations are part of the lab experience. You should record pertinent aspects of such demonstrations, including a schematic diagram of the apparatus, a brief description of the principle of operation, and any especially significant quantitative information. In some courses, each student is required to prepare a report outline or to list the objectives of each experiment in the research notebook.

Recording general information

Each student must record all general information for every laboratory in his or her lab notebook. Examples follow: room T, atmospheric P, free stream velocity, heat exchanger configuration, pump speeds, and so on. Also you should identify all important instrumentation used, including model and serial numbers if available. You should routinely identify by name, date, and size the files that hold extensive computer-based data. And you should always date every entry.

Recording specific data

When the data set is brief, every student should record it all. For longer data sets, students should record at least a representative sample, always including the units of measurement. For example, the vapor pressure experiment illustrated in the example page below has only four data points. Every student should record all the temperature and pressure data for such a concise experiment. However, it should not be necessary for every student to record all the data for a more complex experiment. For example, the LDV traverse illustrated in Figure 8.1 of Chapter 2 has 34 pairs of data. For such large data sets, it is acceptable for each student to record a sample of data in the notebook both for practice and for later reference (*e.g.*, to later verify units of measurement).

Long data sets should be kept in your spreadsheet file with a copy distributed to every group member. It is good practice to print out a condensed version of the data block(*s*) and insert this exhibit into your notebook. Since a graph is a valuable aid to insight and analysis, consider inserting a miniature graph of the data into the notebook as well. Transparent tape is acceptable for course work, but in professional practice such exhibits should be permanently affixed to the notebook. If you are allowed to use tape instead of permanent paste, initial over the edge to preclude tampering. The resulting compilation is an invaluable resource, especially for a long duration research project. However, in undergraduate lab courses, students will be allowed to save time by omitting the attachment of these exhibits to their lab notebooks.

Special requirements for patents

Information and data relating to conceptions that may be patented require special consideration. At a minimum, such entries should be read and signed by at least one knowledgeable and reliable witness. The date on which this takes place should always be recorded by the witness, in some free space adjacent to the signature.

Summary

In summary, these simple guidelines should govern your actions as you keep a notebook. Always number the pages to be used. Date every entry. Every individual should record the general data (*e.g.*, room temperature and ambient pressure) for each lab. Every group member should record important observations and at least a representative sample of the routine data along with the verified units of measurement. List the important instrumentation and equipment. A generic description along with commercial identification (*e.g.*, manufacturer, model, and serial number) is best. Routinely identify by name, date, and size the files that hold extensive computer based data. In professional practice, permanently pasting a printout of the data into the notebook is advisable. For large data sets, consider a shrunken version of the printout. Since a graph is a valuable aid, consider including a miniature graph in the notebook. The notebook is an important professional tool, so develop good practices now.

An example notebook page, prepared for a vapor pressure experiment, follows this section. As you review that page, please note that this page exemplifies the three most important features of a research notebook: the pages are numbered, the appropriate notes and data are recorded and dated, and the notebook represents a continuous and contiguous record of your activities.

Finally, it is worth noting that in recent years, electronic lab notebooks have become more popular, particularly in research labs where you might perform undergraduate research. For an overview of electronic lab notebooks (ELNs), please see the next chapter.

Table 16.1 itemizes the minimal guidelines for effective maintenance of the research notebook.

Table 16.1. Guide to maintaining experimental engineering research notebook

Number all pages to be used.	Date all entries.
Use ink for all entries.	Z-out any large areas left blank.
Maintain a continuous and contiguous record.	Record all pertinent information, data, and observations.
Identify all important experimental equipment.	Record all general data for every experiment.
Record at least a sample of detailed data.	Optionally, paste in a spreadsheet block or other listing of computer based data.
Optionally, include a miniature graph of the data.	

SAMPLE PAGE FROM EXPERIMENTAL NOTEBOOK

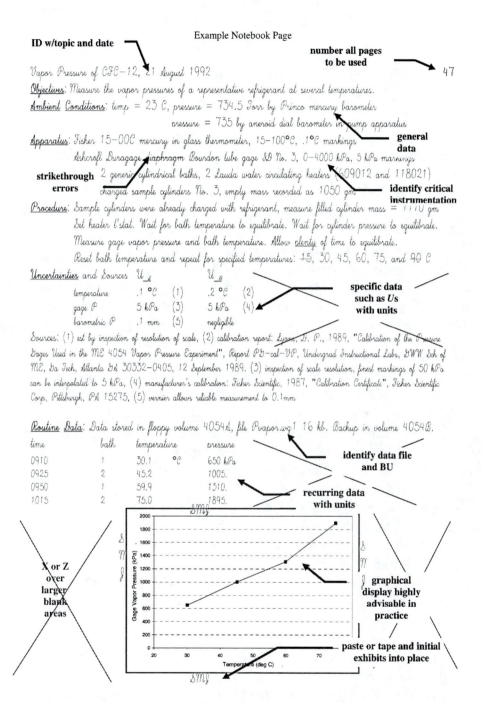

Example Notebook Page

ID w/topic and date

number all pages to be used

47

Vapor Pressure of CFC-12, 21 August 1992

Objectives: Measure the vapor pressures of a representative refrigerant at several temperatures.

Ambient Conditions: temp = 23 C, pressure = 734.5 Torr by Princo mercury barometer

pressure = 735 by aneroid dial barometer in pump apparatus

general data

Apparatus: Fisher 15-00C mercury in glass thermometer, 15-100°C, .1°C markings

Ashcroft Duragage diaphragm Bourdon tube gage SD No. 3, 0-4000 kPa, 5 kPa markings

strikethrough errors

2 generic cylindrical baths, 2 Lauda water circulating heaters (609012 and 118021)

changed sample cylinders No. 3, empty mass recorded as 1050 gm

identify critical instrumentation

Procedure: Sample cylinders were already charged with refrigerant, measure filled cylinder mass = 1110 gm

Set heater t'stat. Wait for bath temperature to equilibrate. Wait for cylinder pressure to equilibrate.

Measure gage vapor pressure and bath temperature. Allow plenty of time to equilibrate.

Reset bath temperature and repeat for specified temperatures: 15, 30, 45, 60, 75, and 90 C

Uncertainties and Sources U_A   U_B

temperature	*.1 °C*	*(1)*	*.2 °C*	*(2)*
gage P	*5 kPa*	*(3)*	*5 kPa*	*(4)*
barometric P	*.1 mm*	*(5)*	*negligible*	

specific data such as *U*s with units

Sources: (1) est by inspection of resolution of scale, (2) calibration report: Lyons, D. P., 1989, "Calibration of the Pressure Gages Used in the ME 4054 Vapor Pressure Experiment", Report PG-cal-VP, Undergrad Instructional Labs, GWW Sch of ME, Ga Tech, Atlanta GA 30332-0405, 12 September 1989. (3) inspection of scale resolution, finest markings of 50 kPa can be interpolated to 5 kPa, (4) manufacturer's calibration: Fisher Scientific, 1987, "Calibration Certificate", Fisher Scientific Corp, Pittsburgh, PA 15275, (5) vernier allows reliable measurement to 0.1mm

Routine Data: Data stored in floppy volume 4054A, file Pvapor.wg1 16 kb. Backup in volume 4054B.

time	*bath*	*temperature*	*pressure*
0910	*1*	*30.1 °C*	*650 kPa*
0925	*2*	*45.2*	*1005.*
0950	*1*	*59.9*	*1310.*
1015	*2*	*75.0*	*1895.*

identify data file and BU

recurring data with units

X or Z over larger blank areas

graphical display highly advisable in practice

paste or tape and initial exhibits into place

In this example page from a laboratory notebook, note that page number, general information, and representative data are entered.

CHAPTER 2.17

COMPUTER GUIDELINES AND COMMON ELECTRONIC RESOURCES

Introduction

Most students will do at least some of their laboratory work on computers provided in a laboratory or library cluster or on the separate computers distributed through the lab areas. These computers should have adequate hardware and software for your anticipated needs. However, the installations on these public computers are likely to differ somewhat from those on the computers you use elsewhere; it is important for you to keep track of the software versions you encounter on different computers so you can anticipate and prevent compatibility problems.

This chapter covers guidelines for computer usage and also gives an overview of some common electronic resources that may be available to you as an undergraduate, including cloud storage, learning management systems, and electronic lab notebooks.

Public computer usage

In an instructional computer cluster, you may be required to use a designated directory for temporary storage of your document and data files; however, you should always copy your files to your own media, such as a flash drive, or to a "cloud" server, and back this up to your personal laptop or home computer for permanent storage. You should assume that the temporary directories in public student clusters will be periodically purged. Consider the following guidelines when using any public computer cluster.

1. Use your own flash drive—or other data storage media—during your data processing. Back up your work often. Students are responsible for loss of data due to failure to preserve backup copies.

2. No unauthorized programs are permitted to be run on the computers located in public clusters.

3. You should not alter the program preferences, options, or other software settings on any of the applications, such as the word processor or spreadsheet programs. Please leave the software in the same configuration as you found it to avoid confusing or delaying the next user.

4. It is unlawful and dishonest to duplicate copyrighted material, so do not copy any applications software from the public computers.

5. If a system malfunctions, you should inform the user assistant, a lab or cluster teaching assistant, the lab supervisor, or a laboratory faculty member as soon as possible.

6. In computer clusters, the hard disks are for permanent applications programs only.

Do not alter, copy, or change any of the program files on the hard disks. Students may use the designated directory for temporary storage of files, but should not rely on these directories for permanent storage of any personal files.

These guidelines should help you to avoid the common problems encountered with public computers.

Other laboratory computers

The same guidelines typically apply for computers other than public cluster computers, such as computers used with or integrated into experiments with the one exception that no personal files should ever be installed on such computers. It is a good practice to scan your media for malware before installing it in such lab computers.

Software usage

Most users will have little difficulty with the programs that are installed in computer clusters, with the possible exception of mismatches with versions installed on other systems that they use.

Required Computer Ownership

Many colleges and universities now require incoming students to bring their own personal laptops for use in their coursework. As policies on required computer ownership differ widely, it is best to check with your department or institution's website before purchasing or leasing a laptop or desktop computer. In particular, take note of any minimum requirements for memory, storage, and security, as well as mandated software and operating systems. Additionally, instructors' policies may differ regarding when and how you are allowed to use your laptop or mobile device in the classroom. Please check the syllabus to see what your instructor's policies are regarding use of electronics during class time.

Cloud Storage

One useful tool for electronic file storage is known as cloud storage, which involves the use of Internet servers to store files such as written documents, spreadsheets, presentations, pictures, and videos. (Although commonly referred to as "the cloud," what this really means is that your files are stored on a remote server that you can access through the Internet.) Some cloud storage is free or very inexpensive and allows you to save or back up your files easily. You can then access your files from anywhere that has an Internet connection. Most of these services enable sharing of data across devices as well, so you can access files both on your home computer and public computers as well as on your phone or tablet using mobile apps.

Cloud storage also facilitates real-time collaboration on documents or presentations, which can be particularly useful in a lab or other class where you are creating a report or presentation as a group. It allows all members of the group to edit work products simultaneously and enables others to view your changes and comments. Sharing reports, presentations, and so on using a cloud server can help ensure that everyone is editing the same version, thus reducing the probability of "version control" problems. It also eliminates the need to constantly email new versions of documents back and forth.

Most cloud storage services offer some amount of storage for free; it is best to check these services' websites to determine how much free storage is currently offered. Some users opt for a paid subscription, which may range from a few dollars a month to hundreds of dollars a year, for much more storage. Some colleges and universities have licenses with these services that allow faculty, students, and staff either limited or unlimited storage capacity on the cloud servers. Check with your college or university to find out whether you have access to this type of free cloud storage.

Commonly used cloud storage services:

Google Drive: This service might be the one you are used to from high school or from other college classes. Google Drive offers some free storage and also is generally compatible with Microsoft Word, PowerPoint, and Excel.

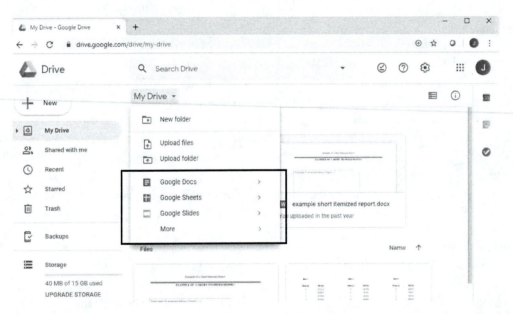

Figure 17.1. This screenshot shows thumbnails of reports, presentations, and spreadsheets stored on a sample Google Drive. The boxed portion shows how you would access Google Docs, Sheets, and Slides under "My Drive."

Google Drive enables storing, sharing, and simultaneous editing of different types of documents: Google Docs for written reports, Google Slides for presentations, and Google Sheets for spreadsheets. Within Google Drive, you may open and share documents and presentations that were created in Microsoft (MS) Office; however, note that some of the formatting from MS Office software might not be preserved in Google Drive. Be aware of the need to proofread and check formatting before final submission.

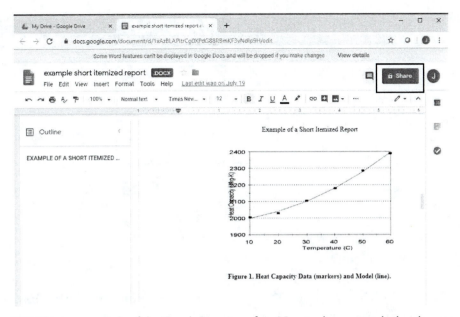

Figure 17.2 This is an example of the Google Docs interface. Note in the upper right-hand corner, there is a box labeled "Share." Clicking on the "Share" box opens a dialog box in which you may enter the email addresses of others with whom you would like to share your document.

Microsoft OneDrive: Similarly to Google Drive, Microsoft OneDrive offers storage and sharing of various types of documents, as shown below.

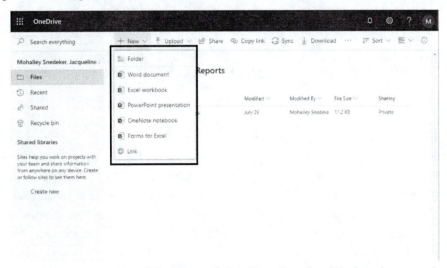

Figure 17.3. This is a screenshot of the Microsoft OneDrive Interface. The boxed portion shows the types of documents that can be uploaded and shared.

You can choose to share entire folders or just individual documents with other students by emailing them a link, as shown in the boxed portion below.

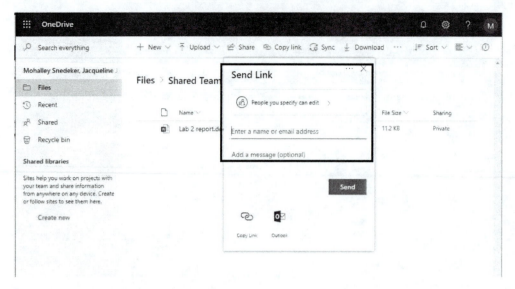

Figure 17.4 This screenshot of Microsoft OneDrive shows how documents may be shared between different users using the "Send Link" feature.

Because it is part of the Microsoft suite of products, OneDrive tends to better preserve document formatting if you are using Microsoft Office to compose your reports or presentations. However, the amount of free storage offered by various products might differ, and you might prefer the features or layout of one service over another. When working on a group project, be sure to consult with your instructor and with the other members of your group before choosing which service to use for storing and sharing documents.

There are many other similar services. Two common ones are Dropbox.com and Box.com. Although these are slightly less commonly used for undergraduate courses, they are still popular and offer similar functionality to Google Drive and OneDrive.

Learning Management Systems

Many institutions use an online learning management system (LMS), which provides a means for instructors to post syllabi, quizzes, assignments, grades, and announcements for their courses. For practical purposes, the LMS essentially takes the form of a hosted website for the course; access to the website is usually restricted to course participants such as students, instructors, and teaching assistants. Many LMSs also offer online gradebooks, discussion boards, collaboration tools, real-time conferencing, and other helpful features. You may be required to submit assignments or post questions to the LMS; be sure to check your instructor's syllabus for specific instructions.

Common learning management systems in higher education include Canvas, Moodle, Blackboard, and D2L Brightspace, among many others.

Below is an example of a Canvas course home page for a laboratory course.

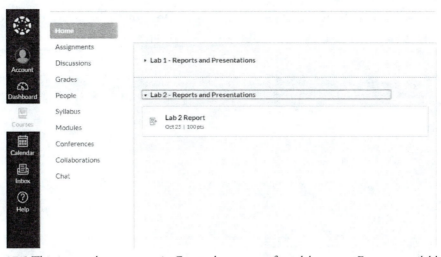

Figure 17.5 This image shows a generic Canvas home page for a lab course. Features available within Canvas include a course calendar, syllabus, modules with assignments, and online conference and collaboration tools.

Electronic Laboratory Notebooks

All undergraduate lab classes will require you to keep a lab notebook to record your data, as discussed in the previous chapter. Keeping a neat and well-organized notebook (including dates and signatures of teaching assistants or lab instructors) helps to ensure that your data are accurate, verifiable, and easily accessible when writing a report or making a presentation about your findings.

Although physical lab notebooks are still the most common requirement for undergraduates, some institutions have begun implementing electronic lab notebooks (ELNs) in lab courses and elsewhere. Electronic notebooks offer several advantages over the traditional physical notebook. With an ELN, students and instructors can more easily record, access, and share data, thus easing collaboration. ELNs may be either "cloud-hosted," meaning they are stored securely on a public or private server on the internet, or locally hosted on a server belonging to your department or institution. However, concerns about data security and tampering, as well as the cost and learning curve of adopting ELNs, still remain a barrier.

Most ELNs offer features such as time-stamping, electronic signatures, and tamper-proofing, which help ensure the integrity of the data shared. If you perform undergraduate research as part of a team of students and researchers, you may be asked to use an ELN to record and share data. Many ELNs offer a certain amount of free storage, but some do require a subscription.

There are many electronic notebook options available, ranging from generic note-taking software such as OneNote and EverNote to more specialized ELNs such as LabArchives or Benchling, among many others. You should check your course syllabus or ask your instructor to see if you are required (or permitted) to use an electronic lab notebook.

Below is an example of the OneNote interface. OneNote is included as part of the Microsoft Office suite, making it a no-cost or low-cost option for most departments and researchers.

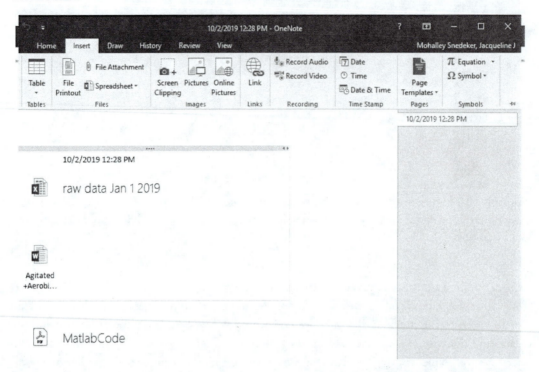

Figure 17.6 This is a screenshot of the OneNote interface. The image shows the Insert menu within OneNote, which allows users to insert spreadsheets and other files as well as pictures, timestamps, and equations into a lab notebook. You may also record and insert audio and video into your notebook using a tablet or other mobile device that has OneNote enabled.

CHAPTER **2.18**

VIRTUAL COMMUNICATION PLATFORMS

Introduction

Classroom and office environments have recently changed drastically with the move to online courses and virtual meetings. You may have had to use your university's Learning Management System (LMS, such as Canvas or Moodle) for classwork more than ever before. Although you likely had ample experience with your LMS before the transition, you may now be required to learn additional new software designed for virtual collaboration and meetings. The good news is that learning these platforms will better prepare you for your first job or internship—as they are also used in industry. This chapter will focus on two of the more widely used platforms: Zoom and Microsoft Teams.

Video Conferencing Etiquette

Just as in person meetings require professionalism on your part, so does shifting to the virtual world. As tempting as it is to attend a meeting in pajamas, you will want to take virtual meetings as seriously as you do in person meetings. Set yourself up for success by following some etiquette guidelines for video conferencing.

If you will take meetings from home, identify a location in your house where you can sit at a table and take notes. Avoid taking meetings from your couch, lounge chair, or bed! If you do not have dedicated desk space, your kitchen table is a good option. Ensure that the video background is free of anything that would be considered unprofessional. Select a quiet location and mute yourself when not speaking—you want to avoid a noisy roommate or barking dog interrupting the meeting. Other items to consider are no different than for face-to-face meetings: dress professionally, arrive on time, and stay on task.

Profile photo

You will likely use video conferencing platforms for a wide range of settings including courses, meetings with classmates and faculty, and interviews with potential employers. Therefore, your profile should be a professional representation of yourself and should look the part. One of the first steps to setting up a profile is to upload a photo. The same holds true when selecting a photo for your LinkedIn profile discussed in Chapter 3.7. Have a friend or family member take a few photos of you in professional attire and simple background.

Guide to Zoom

The Zoom teleconferencing software (https://zoom.us/) is a powerful video platform which allows you to schedule and attend virtual meetings. The software is free but has some limitations such as a 40-minute time restriction on meetings and maximum of 100 participants. The upgraded version removes these restrictions.

Setting Up Your Profile

When setting up a Zoom account, you have the option to create an account using your personal email or university credentials. As mentioned above, a free personal account will come with basic features unless you choose to upgrade, while a university purchased license will allow access to all features. Figure 18.1 shows the sign-up page using either your personal email address or your university credentials using a single sign-on (SSO).

After signing up for your Zoom account, you can begin personalizing your profile and updating basic settings in the Profile page, Figure 18.2. Important items on the Profile page include your profile photo, display name and pronouns, and company (or university). Continue scrolling to update your time zone and personal meeting room ID (available with premium subscriptions only).

After updating your basic settings, you should download (Figure 18.3) the more user-friendly Zoom Client Dashboard shown in Figure 18.4. Once you have installed the Zoom Client Dashboard, you have quick access to settings by clicking on your profile photo, Figure 18.5.

Figures 18.6-18.8 show additional video and recording settings which will be of interest to you.

Create a free Zoom account with your personal email address.

Use SSO to sign-up with your university credentials.

Figure 18.1. Creating a Zoom account

Upload a professional profile picture by clicking on the white pencil.

Continue scrolling to complete additional basic profile information.

Edit general display information.

Set your time zone.

Figure 18.2 Basic profile template after signing up for a Zoom account

Download the Zoom Client Dashboard under the Resources tab.

Figure 18.3 Zoom Client download

221

Figure 18.4 Zoom Client Dashboard

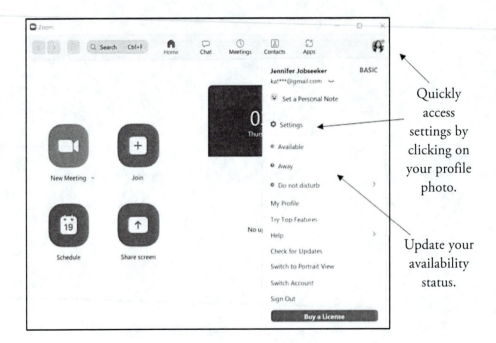

Figure 18.5 Quick settings within the client dashboard

Confirm your room lighting, video framing, and background before joining your first meeting.

Try additional features to optimize your video.

Figure 18.6 Video settings

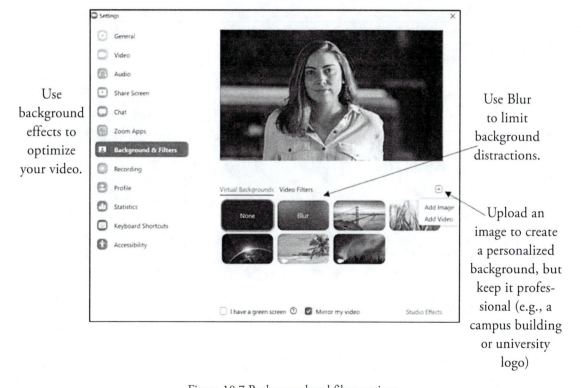

Use background effects to optimize your video.

Use Blur to limit background distractions.

Upload an image to create a personalized background, but keep it professional (e.g., a campus building or university logo)

Figure 18.7 Background and filter settings

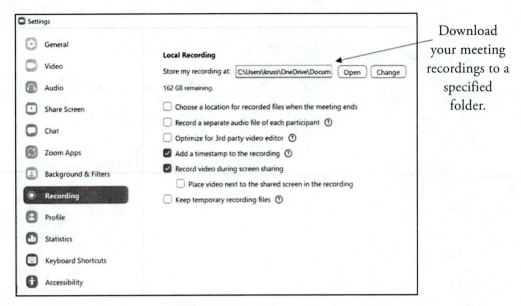

Download your meeting recordings to a specified folder.

Figure 18.8 Recording settings

Meeting in Zoom

The Zoom Client Dashboard presents four options for meetings. Here you can simply start a new meeting, join someone else's meeting, schedule a meeting, or start a meeting with your shared screen. Each of these options is shown in Figure 18.9. You can additionally join Zoom meetings from email links and scheduled calendar invites.

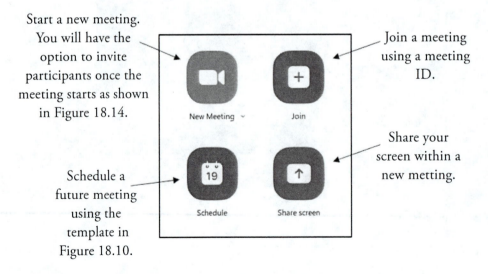

Start a new meeting. You will have the option to invite participants once the meeting starts as shown in Figure 18.14.

Join a meeting using a meeting ID.

Schedule a future meeting using the template in Figure 18.10.

Share your screen within a new metting.

Figure 18.9 Meeting options on the Zoom Client Dashboard

Scheduling a Zoom Meeting

Zoom allows you to schedule a future meeting, specify meeting settings, invite attendees, and create a calendar invite all in a few simple steps. Figures 18.10-18.12 outline the process for scheduling a meeting once you have selected the Schedule option on the Client Dashboard.

Rename your meeting so attendees will quickly know the topic.

Select a Recurring meeting to populate multiple calendar invites.

Generate a new Meeting ID or use your Personal Meeting Room.

Use a Waiting Room if attendees will join at different times.

Create a calendar invite in chosen platform. Figure 18.11 shows the setup for Outlook.

Select meeting date and duration.

Confirm time zone selected when originally setting up your Zoom profile, Figure 18.2.

Require a passcode for added security.

Select whether you and your attendees will automatically enter the meeting with video on or off. Selecting Off gives attendees the option to turn video off or keep on.

Close this window and open calendar options in chosen platform.

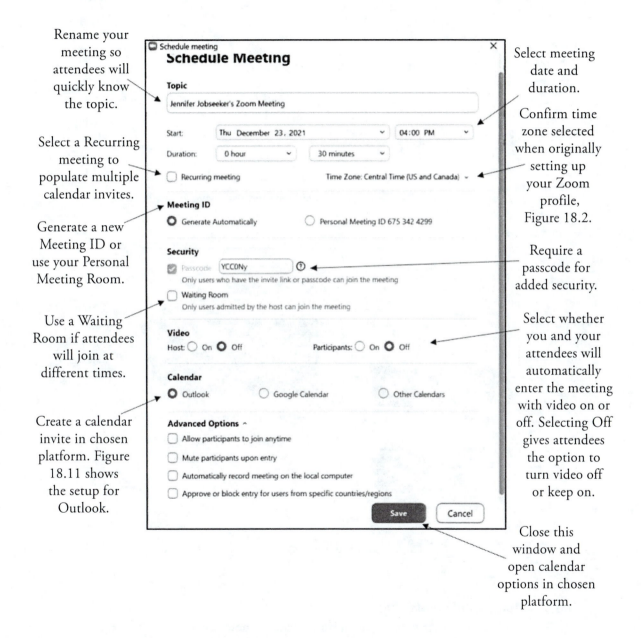

Figure 18.10 Scheduling an upcoming meeting

Meeting details will transfer over from the Zoom setup.

Add required and optional attendees by email address and Send.

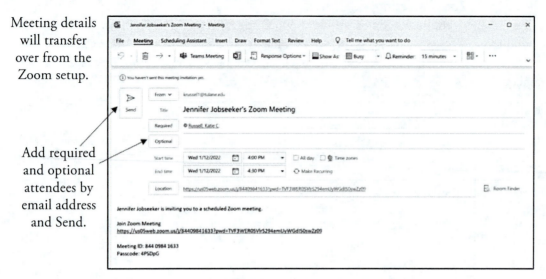

Figure 18.11 Setting up an Outlook Calendar invite

Meeting details will appear in your Zoom Client Dashboard.

Share the meeting with additional attendees by copying the invitation.

Figure 18.12 Sharing a scheduled meeting from the Client Dashboard

Zoom Features within a Meeting

Whether you are the meeting host or an attendee, Zoom has numerous features to enhance your meeting experience. Figures 18.13-18.16 demonstrate many of the most useful settings and features.

Click on the circumflexes for quick access to additional settings.

The ellipsis provides quick access to display settings.

Quickly change your profile picture or display name. Note that your profile picture will appear when your video is turned off.

Toggle audio and video on and off.

Exit the meeting.

Figure 18.13 Basic features within a Zoom meeting

View meeting participants and invite additional attendees.

Share your screen with meeting participants.

Create Breakout Rooms.

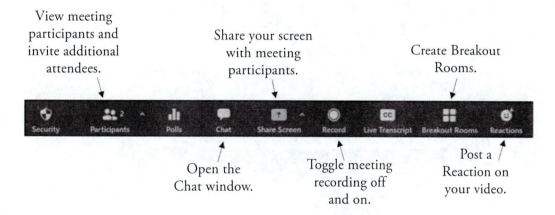

Open the Chat window.

Toggle meeting recording off and on.

Post a Reaction on your video.

Figure 18.14 Features available within a Zoom meeting

A light green border appears while screen share is active.

Select to Stop, Start a new, or Pause screen share.

Open the Annotate dashboard.

Toggle between view options for meeting participants.

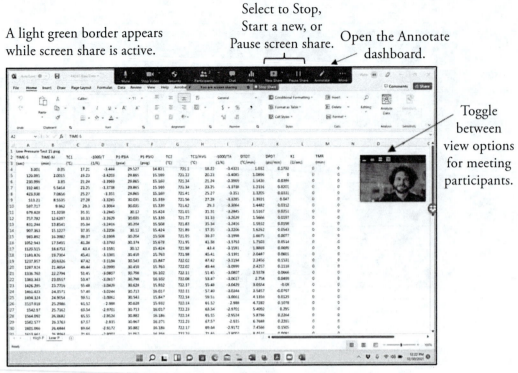

Figure 18.15 Screen sharing in Zoom

Format lets you change colors.

Tools available for annotating on a shared screen.

Save a copy of the annotated document.

The draw tool lets you select multiple shapes.

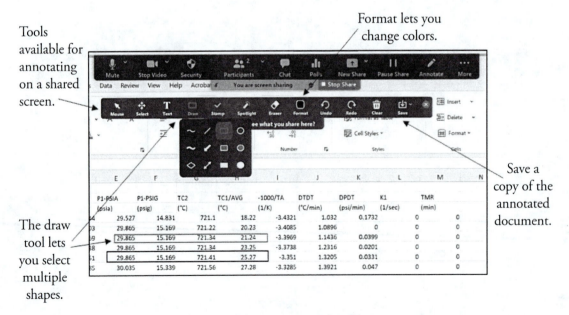

Figure 18.16 Annotation tools in Zoom

Guide to Microsoft Teams

Microsoft Teams (https://www.microsoft.com/en-us/microsoft-teams/group-chat-software) is a communication platform that brings groups of people together for videoconferencing, chat, and file storage. Although a free version is available, Teams is part of Microsoft 365 so anyone with a subscription to the Microsoft suite has full access to the platform. You will have the option to download Teams using your personal email address, or your work or school address shown in Figure 18.17.

Joining and creating Teams

Aptly named, the Microsoft Teams platform groups projects into "teams." You can create a team for any situation where you want to bring a group of people together. Some examples include a class project team, extracurricular organization, or study group. Once you join a team, you have access to the users, chats, and files associated with that team.

Figure 18.18 shows the Teams dashboard for an account with three active teams. Selecting to create a new team will walk you through a set of prompts such as team type, name, description, privacy settings (public vs. private team), and team members.

To access the full version of Teams, create an account using the email associated with you Microsoft 365 subscription.

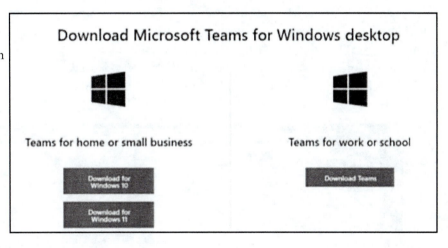

Figure 18.17 Downloading Microsoft Teams

Private chat with one or more team members.

Locate active teams in the Teams Dashboard.

Access to your Teams Calendar.

Join or create a new team.

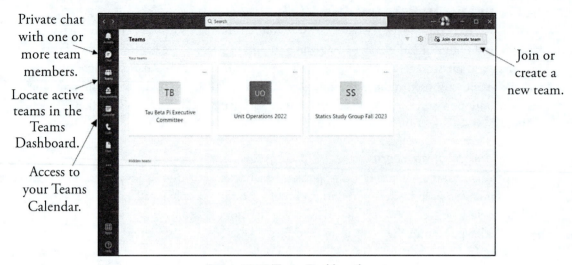

Figure 18.18 Teams Dashboard

Team Organization

Figure 18.19 shows a newly created team titled American Institute of Chemical Engineers. The ellipsis next to the team name provides options for managing and editing the team. Each team is organized by a series of channels and additional channels can also be added here. Note that a newly created team will only have a single channel (General). In Figure 18.20, a new channel (Weekly Announcements) has been created and the channel settings are shown.

Quick access to team settings.

Add a channel to the team.

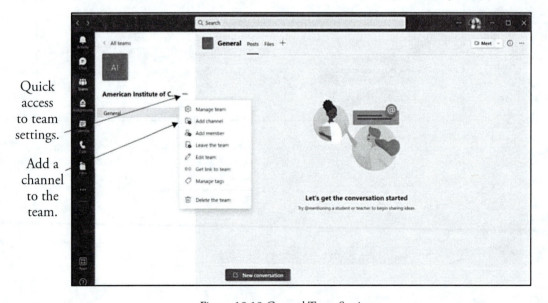

Figure 18.19 General Team Settings

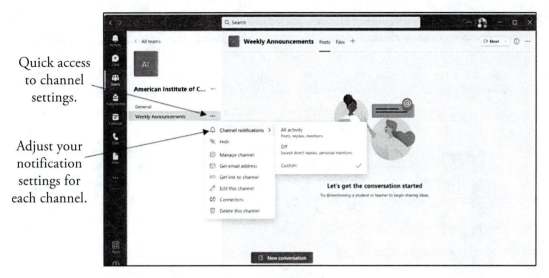

Quick access to channel settings.

Adjust your notification settings for each channel.

Figure 18.20 Channel Settings

Meeting in Microsoft Teams

The Teams platform makes it simple to attend a meeting with members of a specific team. Within a channel, you can simply select to Meet now, or Schedule a meeting, shown in Figure 18.21. Additionally, you have the option to join a scheduled meeting from a confirmation email or a scheduled calendar invite, Figure 18.22.

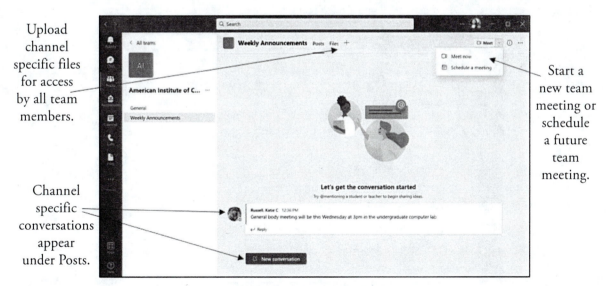

Upload channel specific files for access by all team members.

Start a new team meeting or schedule a future team meeting.

Channel specific conversations appear under Posts.

Figure 18.21 Meeting in Teams

Scheduling a Teams Meeting

Teams allows you to schedule a future meeting, specify meeting settings, invite attendees, and create a calendar invite all in a few simple steps. Figure 18.23 outlines the process for scheduling a meeting once you have selected the Schedule option either from within a channel (Figure 18.21) or from the Calendar (Figure 18.22). If you specify a channel location for the meeting, the meeting information will display in the channel specific posts, as shown in Figure 18.24.

Scheduled meetings can be accessed through the Calendar entry.

Figure 18.22 Teams Calendar

Name your meeting so attendees will quickly know the topic.

Add email addresses of required and optional attendees.

Assign the meeting to a specific channel.

Select meeting date and duration.

Select a recurring meeting to populate multiple calendar invites.

Add meeting details. Consider adding an agenda here.

Figure 18.23 Scheduling a Teams meeting

Figure 18.24 Meeting details populated into specified channel

Features available within a Teams Meeting

Whether you are the meeting host or an attendee, Teams has numerous features to enhance your meeting experience. Figure 18.25 displays the most useful settings and features.

View meeting participants and invite additional attendees.

Post a Reaction on your video.

Share your screen with meeting participants.

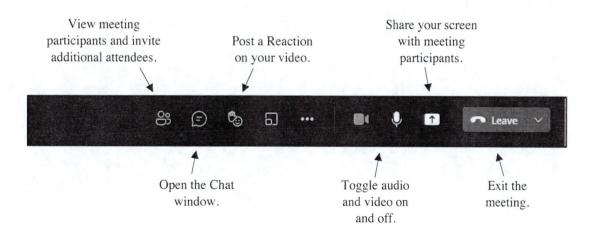

Open the Chat window.

Toggle audio and video on and off.

Exit the meeting.

Figure 18.25 Features available within a Teams meeting

CHAPTER 2.19

SIGNIFICANT DIGITS AND UNCERTAINTY

Most engineering laboratories involve measurements using various instruments or sensors (e.g. scales, pressure sensors, calipers, tape measures, load cells, strain gauges, etc.). Any measurement from an instrument will have an associated uncertainty. The uncertainty is due to the *precision* of the instrument, the *repeatability* of the instrument, and the *calibration* of the instrument. The precision refers to the lowest increment of measure provided by the instrument and the repeatability refers to whether the instrument provides the same reading with repeated measurements. The instrument is usually calibrated to some standard measure but over time the properties of the sensor may change slightly so that it is no longer "reading correctly." This obviously introduces a bias into all the measurements made with that instrument, and hence is another source of uncertainty.

Some estimate of the uncertainty is always necessary for any measurement to be useful in experimental engineering or design. For example, the uncertainty is needed in order to determine if the measurement agrees with alternative experimental data or a theoretical prediction, and obviously an uncertainty is needed in order to assign a reasonable margin of safety in design.

The estimate of the measurement uncertainty should be part of your experimental work and data analysis. Different disciplines take different approaches to measuring and accounting for experimental uncertainty. For work involving very precise measurements, statistical analysis may be needed to calculate uncertainty. For larger scale testing with more coarse measurements, it may be sufficient to use engineering judgment to estimate the uncertainty. For example, it is well known that many construction materials (e.g. concrete, steel, timber) have strength values that can easily vary by ± 10% or more. An uncertainty in an instrument measurement of ± 1% is negligible compared to this, and may be reported in a more informal way. Consult your instructor for the expectations for your particular discipline. In addition, many statistics textbooks have a discussion of measurement uncertainty.

In writing your report, the measurement uncertainty should be reported or at least addressed when discussing your results. A typical means of reporting uncertainty is to provide the measured value and a "confidence interval." Consider the example shown in Fig. 18.1, which shows a sample on a digital scale. The reading is "3.24 kg". Assume that your analysis of the precision, repeatability, and bias of the scale results in a calculated uncertainty of +- 0.05 kg. Your measurement should be reported as "3.25 +- 0.05 kg." If a statistical analysis was conducted to obtain this uncertainty, the uncertainty usually represents the range within which we expect the actual measured value to lie with a confidence level of 95%.

Associated with the uncertainty is the least *significant digit (LSD)*, which is the smallest digit providing any useful information. In the example above, the LSD is 0.01 kg. Note that both the measurement and the uncertainty interval should have the same LSD.

Because of this intimate relationship between the LSD and the uncertainty, at least an order of magnitude estimate of the uncertainty is needed to identify the LSD. Assuming the uncertainty is known or has been estimated, the LSD can be identified. There should be no digits in a reported value less than the LSD! Given that the LSD of the example scale in Figure18.1 is 0.01 kg, there should never be a reported value like "3.242 kg" (either measured or calculated).

The *significant digits* of a measurement are all non-zero digits and all zeros that are not mere placeholders. Never include trailing insignificant digits in the formal text or tables of reports.

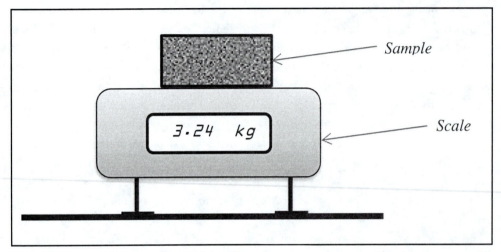

Figure 18.1 – Example measurement.

Table 18.1 shows examples of measurements and their associated lowest significant digits and significant figures. The lowest significant digit for each example has been underlined.

Example 1 is a measurement with 5 significant digits and the LSD in the 10,000th mm place. This is a very precise measurement!

Zeros require special attention so that zeros that are merely place keepers are not erroneously interpreted as significant digits. Report writers must be careful about the occasional trailing ambiguous significant zero or zeros. Most important, all zeros left of the decimal and right of a significant digit must be significant digits.

In Example 2, the 10's place has been identified as the LSD. Therefore, there are only 2 significant digits. On the other hand, in Example 3, the LSD has been identified as the 1's place, and so there are 3 significant digits. Since there is a zero in the 1's place, we can ensure this is clear by including the decimal point.

Example 4 shows that zeros that are "place keepers" are not significant digits. All three of the examples shown have only 2 significant digits. Some editors require that a decimal fraction be written with a leading zero, as in 0.0031 mm. However, note that this leading zero is not to the right of a significant digit, so the measurement still has two significant digits.

Zeros to the right of the decimal and right of any other significant digit must be significant digits. This is illustrated in Example 5.

Example 6 shows some tabulated data. Obviously the 253 mm and 574 mm measurements have 3 significant digits. What about the 310 mm measurement? Although it does not have a decimal point at the end, it would be inferred that it also has 3 significant digits since it is included in the table with these other measurements.

Example 7 shows a measurement in scientific notation. All digits in scientific notation must be significant, therefore the measurement 3.10×102 mm has 3 significant digits.

No general rules on significant digits have been widely adopted or apparently even proposed for statistics. A reasonable rule is to maintain enough significant digits in statistics to calculate a percentage or to make an unambiguous ranking. Example 8 shows two values of R2 that have been calculated from a linear regression. In order to compare, at least 4 significant digits are required. The number of significant digits required to report the results of various statistical tests should be sufficient to compare to the desired confidence level. Example 9 shows that 3 significant digits are required for an unambiguous comparison to the confidence level for this particular p-test.

Table 18.1. Examples of measurements and associated LSD and significant digits.
The LSD in the sample entry is underlined.

Example	Measurement	Number of Significant Digits
1	3.1416 mm	5
2	310 mm	2
3	310. mm	3
4	.0031 mm 0.0031 mm 0.00031 mm	2
5	0.00310	3
6	253 mm 310 mm 574 mm	3
7	3.10×10^2 mm	3
8	a) $R^2 = 0.9999$ b) $R^2 = 0.9995$	4
9	p-value = $0.0495 < 0.0500$	3

LSD and Significant digits for calculated values

Usually values measured during an experiment will be used in equations to calculate various other parameters. For example, the force applied to a steel coupon will be divided by the measured cross-sectional area to obtain the stress. Both the force and the cross-sectional area measurements will have an associated uncertainty. These uncertainties must be combined to obtain the uncertainty for the calculated stress. This uncertainty may be reported along with the calculated stress value, but in the very least it will be used to determine the significant digits.

The mathematical analysis of uncertainty can be used to show how uncertainties should be combined for various calculations. Only the final rules will be presented here. The rules depend on whether the calculation involves a sum, difference, product or quotient:

1) For a sum or difference: The LSD of the sum or difference corresponds to the larger LSD in the two terms of the sum.

2) For a product or quotient: The number of significant digits of the product or quotient is the same as the number of significant digits in the factor with the fewer significant digits.

Example

Table 18.2 shows data obtained from an experiment. The experiment involved measuring mass flow and input and output temperatures. The flow meter was read to the sixth decimal place ("raw m-dot") but it was calibrated ("corr m-dot") to only ± .0003 kg/sec. The thermometer was read to the hundredths decimal place ("raw T-in" and "raw T-out") but was corrected ("corr T-in" and "corr T-out") by a calibration function with uncertainty of .2°C, the tenths place.

The change in temperature ("delta-T") and average temperature ("T-avg") were calculated using the corrected temperature measurements. Finally, the "heat rate", Q, was calculated by:

$$\dot{Q} = \dot{m} C_P \left(T_{out} - T_{in} \right)$$

(18.1)

where $\dot{m}$ is the mass flow, and C_p is the specific heat. The specific heat, C_p, was a given value assumed to vary negligibly over the entire temperature range.

Several significant digit problems can be identified with the reported values in Table 18.2. The corrected mass flow and temperatures have more significant digits than can be justified given the least significant digits in the flow meter (the ten-thousandth place) and thermometer (the tenths place) calibrations.

The reported values have been revised in Table 18.3 to be in accordance with good practice. The raw measured flow rates and temperatures remain as in Table 18.2. However, the corrected values of the flow rates are rounded to the fourth place and the temperatures to the tenths place in accord with their respective calibration uncertainties.

The corrected temperatures are now reported in the more standard units of Kelvin by adding the offset 273.15, which is an exact value and does not move the least significant digit.

Based on the rule for significant digits for sums and differences, the reported temperature difference and average temperature values should have the largest of the LSD of the terms in the sum or difference. The LSD for the corrected temperatures is the tenths place, and hence the temperature difference and average temperatures are reported to the tenths place.

The heat rate calculation involves multiplication of two measured values and an exact value. Based on the rule for products and quotients, the number of significant digits for the calculated value should be identical to the factor with the smallest number of significant digits. The corrected mass flow has only 2 significant digits and hence governs, leaving the computed heat rate with 2 significant digits. Note the first heat rate, "0.60", is reported with the trailing zero because this is in fact a significant digit.

Finally, the formatting of Table 18.3 is an example of professional presentation standards. In Table 18.2, the input temperature is crudely written as "T-in" while in Table 18.3 it is written with subscript and italicized: "T_{in}". Temperatures in Celsius are shown with the degree symbol: "$^{\circ}C$" and the temperature difference with the delta symbol: "ΔT".

Table 18.2. Example of Inattention to Significant Digits and Numerical Format

raw	corr	raw	corr	raw	corr			Cp	heat
m-dot	m-dot	T-in	T-in	T-out	T-out	delta -T	T-avg	at T-avg	rate
kg/sec	kg/sec	C	C	C	C	C-deg	C	kJ/kg-K	kJ/sec
0.006788	0.006584	24.21	24.89	45.3	46.69	21.79	35.79	4.179	0.600
0.008593	0.008335	24.33	25.02	40.23	41.47	16.45	33.24	4.179	0.573
0.008934	0.008666	24.23	24.91	39.8	41.02	16.11	32.97	4.179	0.583
0.00948	0.009196	24.12	24.80	36.84	37.98	13.17	31.39	4.179	0.506

Table 18.3. Example of Good Practice in Significant Digits and Numerical Format

raw mass flow	corr mass flux	raw inlet T	corr inlet T	raw outlet T	corr outlet T	T diff	avg T	Cp	heat rate
$\dot{m}$	$\dot{m}$	T_{in}	T_{in}	T_{out}	T_{out}	ΔT	T_{avg}	at T_{avg}	$\dot{Q}$
kg/sec	kg/sec	$^{\circ}$C	K	$^{\circ}$C	K	K	$^{\circ}$C	kJ/kg·K	kJ/sec
.006788	.0066	24.21	298.0	45.30	319.8	21.8	35.8	4.179	0.60
.008593	.0083	24.33	298.2	40.23	314.6	16.5	33.2	4.179	0.57
.008934	.0087	24.23	298.1	39.80	314.2	16.1	33.0	4.179	0.58
.009480	.0092	24.12	298.0	36.84	311.1	13.2	31.4	4.179	0.51

Part Three:
Writing to Land the Job

CHAPTER 3.1

THE APPLICATION PROCESS

Landing your dream job or internship is hard work. Not only is the application process time-consuming and tedious, but the additional steps required once you make it through the first round require polished soft skills in addition to the technical proficiencies that you have mastered through your studies. This section provides an overview of what to expect from the application process. Details on the specific documents required are presented in subsequent chapters.

Table 1.1. Steps to the application process

Step 1	Step 2	Step 3	Step 4
Apply	Interview	Site Visit	Accept the Offer

STEP 1: APPLYING

Before you can land the interview, you must identify positions of interest to you. There are numerous resources, and you should make use of as many as possible. This section outlines the most common methods and guidelines for making the most of each.

Methods to identify open positions

University career services

Your university's Career Center is the best place to start when searching for an internship or entry-level position. Most Career Centers will have a university specific job search platform open only to their students and alumni. Oftentimes, alumni will reach out to their alma mater to directly select students and arrange for on-campus interviews. Therefore, you should use this resource as the best way to connect with prospective employers—and be one step ahead of other applicants. The most common platform is Handshake, which allows students and alumni to upload resumes, search for jobs and internships, apply for positions, register for career events, and manage an online profile available to potential employers. Other options typically available through your Career Center include career fairs, on-campus interviews, and industry meetups and info sessions. If you have not scheduled an appointment with a career advisor at your university, do so immediately to identify all resources available to you.

LinkedIn

LinkedIn is an excellent reference for identifying available positions in your area. Figure 1.1 shows the job search interface. In addition to simply searching for open positions, you should signup for Job Alerts to receive emails when LinkedIn identifies newly posted positions relevant to you.

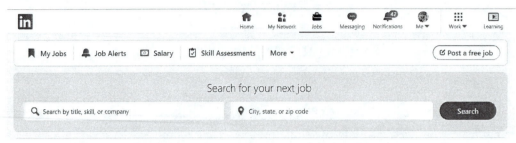

Figure 1.1 The Jobs feature within LinkedIn.

Professional Organization Websites

In addition to networking at the local level, professional organizations typically have career resources available on their national websites. These websites will give you access to jobs specific to your industry and are often only accessible to organization members. Figure 1.2 shows the available features within the American Institute of Chemical Engineers (AIChE) members resources, including finding discipline specific jobs and internships, and networking.

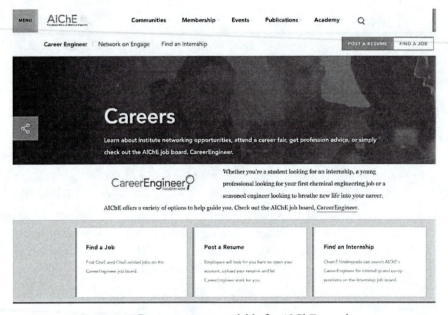

Figure 1.2 Career resources available for AIChE members

Networking

A network is a web of personal and professional contacts that can assist you with your job search and career goals. The majority of positions on the market are not advertised, so the importance of a vibrant network cannot be overstated. During your time as a student, you should work on developing your network using the following avenues:

1. **Program alumni:** Try reaching out to members of your department's Board of Advisors with similar interests and career paths. Another option is to use the search function in LinkedIn to identify alumni with similar interests. Chapter 3.7 concludes with a section on using the search function within LinkedIn to identify potential contacts.

2. **Members of professional organizations:** Joining a professional organization such as the American Institute of Chemical Engineers (AIChE), the Society of Women Engineers (SWE), and the National Society of Black Engineers (NSBE) early in your student career will provide an excellent opportunity to network with professionals within your community. Many organizations also have a Young Professionals Division. Events hosted by professional organizations are a great opportunity to meet new people and learn about what life after college looks like in your field. Consider serving on the executive board of your student organization and working to head-up joint events with the local chapters and young professional members. Such an event will make an excellent bullet point for your early-career resume.

3. **Engineering fraternities:** Many universities support student-oriented engineering societies such as Theta Tau and Sigma Phi Delta. Membership will provide access to alumni, professional development activities, and a lifelong network of fellow engineers.

4. **Family and friends:** Do not underestimate the networking potential within your current set of contacts. Do you have an uncle in the chemical process industry? What about the mother of a close friend? Spend a few minutes brainstorming anyone who could be in your professional network (e.g. family members, family friends, congregation members, advisors, past employers, classmates, coaches, neighbors).

Application documents

Once you have identified available positions of interest to you, you will need to submit a formal application. Documents that may be requested include:

- Resume/CV
- Cover Letter
- Unofficial Transcript
- Supporting Documents

Resumes are covered in more detail in Chapter 3.2, cover letters in 3.3, and CVs in 3.6.

STEP 2. INTERVIEWING

Once you have grabbed the attention of the hiring manager with your application packet, you will be invited to interview. This is typically a multistep process and is often the most intimidating to students—practice makes perfect. This section outlines a typical interview process and tips for success during the process.

The Interview Process

Once you are selected for an interview, you will likely be contacted through email by the hiring manager. Figure 1.3 shows what such an email might look like. If an email response is required, it should be prompt (less than 24 hours), brief, and professional.

Dear Jamie,

I would like to schedule a 30-minute initial phone interview for early next week if possible. This interview will be fairly informal and will determine whether we move you on to the second round of interviews. Below is a link to a calendar sign up. Please pick a time that works best for you. If possible, please schedule for Monday or Tuesday of next week.

https://calendly.com/initialscreeningjamie

Thank you, and I look forward to speaking.

Jordan Smith

Figure 1.3. Example email for a screening interview request

Screening Interview

Oftentimes, companies will conduct an initial round of screening interviews. This type of interview will be brief, usually conducted by phone, and designed to screen numerous candidates to confirm basic qualifications. Expect questions about your educational background, work history, and knowledge of the company. Therefore, you should prepare just as you would for a formal interview using the guidelines in the following section. Note that not all companies will perform an initial screening interview.

Formal Interview

After passing the screening interview, you will be invited for a more detailed interview either in person or virtually. At this point, the company has established that you possess the qualifications necessary to complete the job duties, but passing the next test will require you to truly stand out from the other candidates. Be prepared by demonstrating your knowledge of the company and your fit for the position. Also, do not forget the importance of presentation—dress professionally and arrive early.

Demonstrate your knowledge of the company

You will learn in Chapter 3.3 that a successful cover letter will grab the reader's attention by demonstrating how you are the best fit for the company and position. Similarly, you should arrive to the interview prepared to discuss the same by first thoroughly researching the company to develop a firm grasp of what the company produces, their mission, and, if possible, any background information on your interviewer(s). Use the company website as a starting point, and take note of any recent news stories about the company.

Demonstrate your suitability for the position

The next step to sufficiently prepare for an interview is to compose a set of targeted answers to top interview questions. Some example questions are presented below. Your first attempt at answering these questions will prove challenging. Therefore, you should take the time to formulate a thoughtful response to each which plays up your strengths while downplaying your weaknesses. Write down your answers and revise as needed, then practice often. You will find that once you are truly comfortable with a response, you will be able to tailor it to address a range of potential questions. For example, the same response with only minor modifications may work for any question about teamwork. Check with your university's career center to identify any tools available for you to practice (online learning tools, mock interview with a career coach, etc.).

Categories and sample questions:

Self-Reflection and Life Experiences
- What do you consider your greatest strength?
- What do you consider your greatest weakness?
- What accomplishments to date have given you the most satisfaction?
- What is the biggest mistake you have ever made and what did you learn from it?

Academics
- Why did you select your major?
- What course did you like (or dislike) the most?

Working with Others
- Give an example of a leadership role you have undertaken.
- Have you encountered a conflict in a team project? How did you handle it?

The Position
- What do you know about our company?
- Why are you a good fit for our company?

Work Experience
- Tell me about your work experience.
- How have you demonstrated an ability to work under pressure?

Virtual or phone interviews

Oftentimes, the first round of interviews will be conducted by phone or through a virtual platform like Zoom. Ensure that you have a quiet, distraction free location to conduct the interview. Your career center will likely have quiet rooms for you to reserve for this reason. Even though the interview will be virtual, you should dress neatly and wear the same professional attire that you would wear to an in-person interview.

In some cases, you may be asked to submit a prerecorded video interview. Essentially, there will be a prerecorded video of an employee asking you a question. You will be given a short amount of time (~30 seconds) to gather your thoughts followed by a set recording time (~2 minutes). You cannot rerecord an answer, so jot down your key ideas in the 30 second prep time. Remember to practice sounding upbeat, speaking clearly, and at a reasonable pace.

STEP 3: SITE VISIT

A technical presentation and/or design question is typically the last step prior to a final decision on whether you will be offered the position. Oftentimes, this is completed onsite through an all-expense paid invitation to visit. Figure 1.4 shows an example of what the request may look like.

During your visit or call, you will meet with multiple employees—with and without technical backgrounds. The invitation will outline the specifics of the requested information. Chapter 4.5 provides an Example of a Design Presentation for a Workplace Project, which should be followed in this instance also.

Technical Presentation

You will select a project from coursework or an internship. Be strategic about your chosen project and select one that best demonstrates your problem-solving ability. Remember that you will be presenting to audience members who are not familiar with the project and some may not have technical backgrounds. Therefore, you will want to establish relevance with a solid description of background and goals. Then present your results, discussion, major conclusions, and recommendations moving forward.

Design Question

You will be given access to information regarding a specific problem related to the company. Your goal will be to recommend a solution. Of note in these types of problems is that there is no correct answer. You should design your presentation with a logical explanation of your solution and rationale. Communicate your thoughts explicitly, always restate the problem at hand, and work methodically towards your final selection.

Tips for choosing references

At some point in the interview process, you will likely be asked to provide a list of references who can speak on your character and suitability for the job. Keep your references professional in nature (such as professors and internship supervisors) versus selecting friends and family. You will also want to ask any potential reference for permission to provide their information to your prospective employer. Confirming with your references will ensure that they are comfortable speaking highly of you and are prepared to do so.

Hi Jamie,

As the final step in our interview process, we would like to invite you to an onsite interview where you will have the opportunity to tour the facility and meet our team.

During the visit, we would like for you to give a presentation to a few team members. The presentation should take roughly 1 hour and include the following:

- 30 min presentation of a project
- 15 min design question. Present one or more designs for a fixture that can apply pressure on our lithium metal pouch cell as it cycles. Please see the attached documents: design requirements and spec sheets.
- 15 min Q&A.

Available time slots: Thur 4/30 1- 3:00 pm PST, Fri 5/1 after 1:30 pm PST.

I'll send a calendar invite once you let me know the best 1.5 hour block for your schedule. Additionally, to get a head start on the reference checks, please provide 3 references.

Let us know if you have any questions.

Best regards,
Jim Lopez, PhD
Senior Engineer, BattTech

Figure 1.4. Example Email for Follow-up Interview

STEP 4: ACCEPTING THE OFFER

You will likely receive an email detailing the outcome within 2 weeks of the final interview. The offer should include the amount of time the company expects for a response. Congratulations!

TIPS for evaluating multiple offers

You may run into a situation where you have multiple offers on the table. First off, relax and pat yourself on the back, this is a good problem to have! Keep in mind that there is more to consider than the salary figure when determining the best offer for you. A good method to tackle this important decision is to start by making a list of your most important criteria for future job satisfaction. Items to consider include benefits (retirement funding, paid vacation time, etc.), growth potential within the company, company culture, professional development support (including tuition coverage for advanced education), position expectations, and location. For students deciding between graduate programs, you will most likely be evaluating financial aid, teaching/research expectations, and location. Once you have a clear picture of the criteria that is most important to you, you can weight the pros and cons of each offer and make a decision you can feel confident about.

Chapter 3.2

Guidelines for Resumes

The goal of a resume

Your resume is the first opportunity for a potential employer to learn a bit about you; therefore, an effective resume is one that will leave the reader wanting more. Because hiring managers are oftentimes sifting through hundreds of applications with only a few seconds on each, you will need to ensure that yours makes the cut. This section will provide guidelines for standard resume formatting for students seeking internship or entry-level positions, as well as tips to help you stand out from the pack.

Formatting

Page Layout

You have many options for the overall look of your resume—but you must choose a layout that draws the reader's eye to the most important information. For readability, you want your section headings left-aligned with your section information indented. The level of indentation is up to you but be sure the finished product is pleasing to the eye and easy-to-read. This means plenty of white space and varied spacing, fonts, and formatting. Some choose to use the "hanging heading" layout, where the headings are left-aligned, and the text begins in its own further indented column. This gives less room for additional text but makes the headings standout. A good trick to efficiently align your info is to use tables, demonstrated in Figure 2.1. Other formatting considerations:

- **Spacing:** Single-spaced
- **Font:** Professional, no less than 11 point.
- **Section Titles:** Informative and attention grabbing.

Resume Sections

You learned in Chapter 1 that format is used to describe the sequence of headed sections in a report. Similarly, most resumes are divided into sections arranged in a standard order using the following sections:

1. Heading
2. Education
3. Experience
4. Personalized Sections (e.g. professional memberships, campus involvement, honors, awards, skills)

You may be wondering why the Objective or Personal Statement is not included in this list. Although commonly seen on resumes, the Objective Statement is often a generic statement that is glossed over by the potential employer. Make better use of this valuable space by including additional non-generic personalized sections.

Example 1:

Experience	
Research Assistant	June 2018 – Present
Albert Research Group – Tulane University | *New Orleans, LA*	
• Wrote Python code to determine the Order-Disorder Transition (ODT) points for several polymer blends • Calculated radial distribution functions (RDFs) and analyzed output data to infer system characteristics, such as spacing of lamellar domains	

Experience

Research Assistant   June 2018 – Present

Albert Research Group – Tulane University | *New Orleans, LA*

- Wrote Python code to determine the Order-Disorder Transition (ODT) points for several polymer blends
- Calculated radial distribution functions (RDFs) and analyzed output data to infer system characteristics, such as spacing of lamellar domains

Example 2: Hanging Heading

EXPERIENCE	
May 2019 – Aug. 2020	Engineering Intern
	Shell Oil Company, New Orleans, LA
	• Worked with a multidisciplinary team of engineers to address issues of off-spec and environmental incidents in an industrial-scale distillation column
	• Simulated the column within Aspen HYSYS using historical and real-time data

EXPERIENCE

May 2019 – Aug. 2020   Engineering Intern

                            Shell Oil Company, New Orleans, LA

- Worked with a multidisciplinary team of engineers to address issues of off-spec and environmental incidents in an industrial-scale distillation column
- Simulated the column within Aspen HYSYS using historical and real-time data

Figure 2.1. Layout Examples using Tables for Formatting. Each example includes a) section with borders displayed and b) the final version with "No Borders" selected.

Heading

The heading announces who the resume belongs to, and how the reader can contact you. A proper heading should include:

1. **Your name**
2. **Mailing Address (current and permanent)**
3. **Contact information**

Heading format is ultimately up to the author but be sure that the information is easy to read and neatly organized. Figure 2.2 provides some common heading formats.

Current Address 3958 Toulouse St. New Orleans, LA 70118	**Brittany Nathan** 506.888.5000 | bnathan@tulane.edu linkedin.com/in/bnathan	Permanent Address 5969 Millbrook Dr. Houston, TX 77001

Chase T. Fogg III
8513 State St. • New Orleans, LA 70119 • (338) 896-0000 • cfogg@tulane.edu • linkedin.com/in/cfogg

Harrison "Harry" Levi
975 Rustic Canyon, San Diego, CA 91945
800.333.8000 | harrylevi@tulane.edu | linkedin.com/in/hlevi

Figure 2.2. Examples of Resume Header Format

Education

Your education section is an opportunity to demonstrate what about your academic studies sets you apart from the pack. You have a few options on what to include in this section and remember to use formatting to call attention to highlights.

TIPS for your Education section

- Do not include your high school.
- Include notable courses only if they are unique to the job. For example, you may have taken a series of classes not typically required by your major but relevant to the position (e.g. biotechnology, coding, materials).
- Include any notable projects. This is a good opportunity to insert keywords from the job posting—especially if you have limited experience.

Experience

Your experience section will highlight any work experience you may have. For early career resumes, this section may be slim compared to other sections. That's okay. You can use formatting to your advantage.

EDUCATION   **B.S.E. Electrical Engineering**   May 2024
Minor: Mathematics   GPA: 3.4
Purdue University, West Lafayette, IN

Education

Tulane University, School of Science and Engineering   New Orleans, LA
Bachelor of Science in Chemical Engineering, *Major GPA: 3.5*   May 2023

Honors: Distinguished Scholar Award (top 5% awarded), Dean's List
Relevant Courses: Intro to Biotechnology, Biotechnology Lab, Applied Biochemistry

EDUCATION

B.S.E. Chemical Engineering   May 2022
Minor: Economics   GPA 3.87
Bucknell University, Lewisburg, PA

Notable Coursework
Optimization of chlorinated polyvinyl chloride production – Conducted a sensitivity analysis on flowrates and compositions of raw materials to optimize profits in a simulated industrial plant

Education

Tulane University   New Orleans, LA
B.S. Chemical Engineering   May 2023
GPA: 3.6 Major GPA: 3.9

Relevant Coursework
Chemical Process Design: Conducted literature review and developed plant design strategy for the industrial production of acetic acid. Strong emphasis on process safety, hazards analysis including HAZOP assessment, P&ID analysis, and process optimization.

Unit Operations Laboratory: Planned and performed chemical engineering experimentation on pilot-scale process equipment in a team environment. Strong emphasis on written and oral presentation of technical material; data analysis using Microsoft Excel and Aspen HYSYS; and process hazard screening.

Figure 2.3. Example Education Sections

The amount of experience you have will dictate how much and which items you include here. For example, if you are a graduating senior and have completed multiple internships in your field of interest, you could consider removing the ice cream shop experience—or include it in a separate "Additional Experience" section. If you only have non-relevant experience, that's okay too! Think about how you can use keywords to show that the skills are translatable. For example, did you monitor inventory and place weekly orders, or maybe you managed and trained new employees. Think about all relevant projects you have worked on in both academic and community settings. Another option is to include a Leadership Experience section.

Building your bullets

No matter which experiences you include, you will need to present yourself in the best light by highlighting what sets you apart. You will do this with your bullet points. Here you will learn a technique to build better bullets called the APR technique.

APR (Action + Project/Problem + Results)

Step 1: When considering a specific job, think of a **project** you worked on or a **problem** you solved.

Step 2: Choose an **action verb** to describe the actions you took.

Step 3: Quantify the **results** you achieved.

Table 2.1. Example of the APR Process

Acton Verb:	Simulated
Project/Problem:	Reduction of carbon dioxide emissions in ammonia plant
Results:	Alternative process reduced emissions by 87%
Bullet:	Simulated "green methane" process in Aspen HYSYS® reducing carbon dioxide emission from an industrial scale ammonia plant by 87%.

Acton Verb:	Received
Project/Problem:	Annual internship presentation competition at Intralox
Results:	Won first place
Bullet:	Received the first-place (out of 53 candidates) in the annual internship presentation competition.

Table 2.2. Action Verbs

Analyzed	Coordinated	Engineered	Implemented	Programmed
Built	Created	Evaluated	Improved	Researched
Calculated	Designed	Facilitated	Operated	Simulated
Collaborated	Developed	Generated	Organized	Trained

Additional Sections

Once you have covered your education and experience, you have more freedom in which additional information to include. The following list provides personalized sections that you might consider. Again, always choose the information more relevant to the position you are applying for. To conserve space, you can lump multiple sections together (e.g. Memberships and Certifications):

- Volunteering/Service/Campus Leadership
- Media/Publications
- Professional Memberships and Organizations
- Honors and Awards
- **Skills:** Can be broken into smaller sections (e.g. software, languages, instrumentation, certifications, laboratory, programming)

SKILLS

Software: Microsoft Office • Aspen HYSYS • Adobe Illustrator

Programming languages: C++ • MATLAB

Instrumentation: Flowcytometry • Mass spectroscopy • Vapor diffusion crystallization

---

Figure 2.4. Sample formatting for Skills section

Jennifer G. Jobseeker

1234 Holly Lane, New Orleans, LA 70118

(504) 500-000 | jjobseeker@tulane.edu | www.linkedin.com/in/jennifer-jobseeker

EDUCATION

Bachelor of Science, Engineering (Chemical Engineering)   Expected Completion: May 2023
Tulane University, New Orleans, LA   GPA: 3.67

Study Abroad: University of Auckland, New Zealand   Spring 2021

Academic Honors
Dean's List (Fall 2019-2020, Fall 2022), Tulane Presidential Award, INEOS Student Scholar

Notable Coursework
Chemical Process Design: Designed, simulated, and optimized a process for the industrial production of isopropanol—considering process safety, economic viability, and environmental concerns.

Unit Operations Laboratory: Worked with a team to plan and perform chemical engineering experimentation on pilot-scale process equipment. Strong emphasis on written and oral presentation of technical material, data analysis, and Aspen HYSYS.

RELEVANT EXPERIENCE

Chemical Engineering Intern   August 2022 – December 2022
Shell Oil Company, New Orleans, LA
- Optimized the waste system in a Chlor-Alkali unit to increase production allowance.
- Devised a pre-treatment process which decreased finishing treatment time form 24 hours to 18.6 hours, and increased production capacity by 17%.

Distillation Intern   June 2021 – August 2021
NOLA Spirits Distillery, New Orleans, LA
- Maintained and operated large-scale process equipment for the commercial production of ethanol
- Collected and analyzed product specs to optimize production and monitor quality control.

CAMPUS INVOLVEMENT

Engineers Without Borders, Tulane Chapter   Fall 2021 – present
- Worked with a team to design and build a water purification system for a remote Ecuadorian village.
- Organized the silent auction for the Spring 2022 annual fundraiser.

SKILLS AND CERTIFICATIONS

Software: Aspen HYSYS, Aspen Properties, Aspen Plus, Promax, Microsoft Office
Programming: Excel VBA, MATLAB, LabVIEW
Languages: French (proficient), Korean (basic)
Certifications: Safety and Chemical Engineering Education (SAChE), Level 1-3

Figure 2.5. Example of an Early-Career Resume

CHAPTER 3.3

GUIDELINES FOR COVER LETTERS

The goal of a cover letter

The cover letter is a one-page document presented with your resume. It should further expand on aspects of your resume, not serve as a duplicate, and is an opportunity to fill in the gaps and insert a bit more personality into your application.

A successful cover letter will grab the reader's attention by demonstrating how you are the best fit for their company and position. Therefore, you will want to demonstrate specific qualifications that apply directly to the position. Conduct significant research on both the company and the position you are applying for. The best place to start is the job posting where you should identify specific key words to align with your experience. You should also thoroughly research the company's website to understand how your values align with company culture.

Cover Letter Sections

You have a few options for the layout of text within your cover letter. The key is to choose a layout that draws the reader's eye to the most important information. You will use your text to make the best argument for the position, but the cover letter formatting must follow some basic rules. Below is an outline of the necessary elements to include.

General Formatting

Cover letters should use standard business letter format in addition to the following:
- **Spacing:** Single-spaced, double spaced between components
- **Font:** professional, no less than 11 point
- **Margins:** standard (1-1.5")
- **Paragraphs:** Do not indent the first line and avoid dense paragraphs

Heading

Your cover letter will be presented alongside your resume; therefore, your heading should match the heading chosen for your resume.

Front information

The following should be included after the heading and before the main text.
- **Date:** date of application packet submittal.
- **Name, title, and address of recipient:** If possible, try to identify the person who will be reviewing your application packet. This may be in the job description or you may

be submitting to a contact that you know personally through networking. If you do not have a specific contact, include the general company address.

- **Salutation:** Professionally address the recipient.

First paragraph

States the purpose of the letter and how you found out about the position. Establish an enthusiastic tone and try to include something to catch the reader's attention. Do you have a mutual contact? Something that truly sets you apart, or something you specifically admire about the company? You will also state which opening you are inquiring about.

Body

Here is where you sell yourself. The goal with this section is to demonstrate why you are the best fit for the position. Start by highlighting keywords in the job posting and brainstorm how the items in your resume have prepared you with those skills. The main body of a cover letter is typically broken up into one of two options.

- **Two Paragraphs:** One paragraph elaborating on your education and one on your work experience.
- **Bullets:** Lists your key qualifications with supporting information.

Closing paragraph

The final paragraph is where you will urge your potential employer to get in touch. You can choose to announce that a resume is enclosed for further review. Keeping with the tone of the entire letter, you will want to show your enthusiasm and interest for the position.

Closing Information

In line with standard business formatting, you will end your letter with the following items:

- **Complimentary close:** Sincerely, Respectfully, Best regards are all professional options.
- **Signature block:** Leave 4 blank lines followed by your typed name. Insert your signature above your name.

Proofreading

A potential employer will form their first impression based on your resume and cover letter. Therefore, it is imperative that these documents are polished and free of spelling and grammatical errors. In addition to proofreading during each round of edits, you should ask a few people to review your packet. Consider family members, friends, professors, and university career center advisors.

Chase T. Fogg III

8513 State St. • New Orleans, LA 70119 • (338) 896-0000 • cfogg@tulane.edu • linkedin.com/in/cfogg

15 October 2022

Ms. Jessie Irving
Lead Process Engineer
Moonlight Brewing
567 Hickory Avenue
Boston, MA 02210

Dear Ms. Irving:

I have admired Moonlight Brewing's approach to the beverage production industry since first learning about your transition to renewable energy in 2020; therefore, I was thrilled to see the opening for an entry-level Production Engineer. I am a senior chemical engineering student with the goal to merge my experience in the food and beverage industry with my passion for sustainability. Many of my core strengths, which I've outlined below, align directly with Moonlight Brewing's culture and values:

- **Process design experience in the food and beverage industry** developed at the USDA through start-up and optimization of a pilot-scale sugarcane processing plant.

- **Passion for sustainability** shown by my LEED Green Associate Certification, involvement with Tulane University's Green Club, and commitment to practicing a plant-based diet.

- **Project management experience** through creation of a Process Information Management System with Crescent Consulting and through testing of a new processing aid at the USDA.

- **Strong engineering and biotechnology background** demonstrated by coursework in Biotechnology, Biochemistry, Reactor Kinetics, and Process Design.

- **Appreciation of process safety** through certification in the AIChE Safety and Chemical Engineering Education Program and participation in Process Hazard Analyses (PHAs) at Crescent Consulting.

I am enthusiastic about the opportunity to contribute my skills to Moonlight Brewing. My resume is enclosed for your review, and I am available to answer any questions you may have by phone or email.

Thank you for your time and consideration, and I look forward to hearing from you.

Sincerely,

Chase T. Fogg III

Chase T. Fogg III

Figure 3.1. Example of an early-career cover letter

CHAPTER 3.4

GUIDE TO EMAIL EXCHANGES

Networking Emails

You learned the importance of establishing a robust network in Chapter 3.1. Figure 4.1 provides an example networking email a student might use when reaching out to potential contacts. You should adapt your own email to your personal interests and goals, as well as establish relevance to the addressee.

Dear Mr. Gordon,

I am a senior chemical engineering student at Tulane University graduating in May 2022. I am reaching out because I have great interest in working in the food and beverage industry, and I noticed on LinkedIn that you also studied at Tulane.

I recently applied to a position at General Mills, and I saw that you have worked there for almost 20 years; congratulations on the milestone! Considering your experience with the company, I wanted to get your perspective on the culture there and your experience in the industry.

Would you have time for a phone call to discuss further? Any advice that you may have for a student looking to enter the food and beverage industry would be insightful.

Thank you for your time,
Jessica Green
Chemical and Biomolecular Engineering, Economics minor
Tulane University 2022
jgreen5@tulane.edu
www.linkedin.com/in/jessicagreentulane

Figure 4.1. Sample Networking Email

Thank You Letters

Thank you letters are important to maintaining a viable network. You should write thank you letters in two situations. The first would be to anyone that has helped you (e.g. a professor who wrote a letter of recommendation, someone who you spoke with on LinkedIn, a family friend that provided a contact). The second is a follow-up to a job interview. Below are some tips for an effective thank you letter.

- Be professional and personal
- Demonstrate good communication skills
- If following up an interview, send within 48 hours
- Keep it brief
- Hand-written is more memorable and should be used if there is no time constraint. Email is best if quick turnover is necessary.

A thank you letter which follows up an interview is also a good time to fill in the blanks on the interview. Did you forget to mention an item that aligns well with a particular interview question? If so, use the email to add additional thoughts.

Dear Ms. Bromberg,

I'm writing to thank you for the interview for the electrical engineering position with Honeywell. Speaking with you confirmed my positive impressions of your company and how the culture at Honeywell directly aligns with my interests and career goals.

Please feel free to contact me if you need any additional information. Thank you for your consideration, and I look forward to hearing from you.

Thank you for your time,
Sienna Lewis
Electrical Engineering
University of New Orleans, 2023
Slewis4@uno.edu
www.linkedin.com/in/slewis

Figure 4.2. Sample Thank You Letter or Email

Follow-up Emails

You should always conclude an interview by politely asking when a decision will be made. If you have not heard back within that time frame, you can email the interviewer to request an update. Be sure to be brief and polite.

Dear Ms. Bromberg,

I wanted to follow-up to our interview on November 8th. You mentioned that you expected to decide in about 2 weeks, so I wanted to inquire if the position has been filled.

Thank you,
Sienna Lewis
Electrical Engineering
University of New Orleans, 2022
Slewis4@uno.edu
www.linkedin.com/in/slewis

Figure 4.3. Sample Follow-up Email

Requests for Undergraduate Research

If your undergraduate career is preparing you for graduate studies, you should identify faculty members whose research labs you might want to join. The following is a sample of what a request for additional information about research opportunities may look like.

Dear Dr. Cook,

I am a sophomore majoring in chemical engineering and am interested in conducting undergraduate research during my time at FSU. I was researching the lab groups in our program, and your research stood out to me. Your lab's focus on mesenchymal stem cells and their use in regenerative medicine is a topic that I have always been fascinated with and I would love to learn more.

Would you have time for a brief meeting to further discuss the research in your lab and any opportunities available to undergraduate students? If so, I am free from 2-5pm on Tuesdays and Thursdays, or 9-11am on Wednesdays. Let me know if any of these times would be convenient for you.

I look forward to hearing from you, and thank you for your time.

Mindy Toriello
Florida State University, Class of 2025
mtoriello9@fsu.edu
www.linkedin.com/in/mtoriello

Figure 4.4. Sample Email to Request Undergraduate Research

CHAPTER 3.5

THE GRADUATE SCHOOL APPLICATION PROCESS

Just as landing a dream job in industry takes time and devotion, so does applying to graduate school. This section provides an overview of what to expect from the application process, and details of specific application components follow in subsequent sections.

Table 5.1. Steps for the Graduate School Application Process

Step 1	Step 2	Step 3	Step 4	Step 5
Identify Programs of Interest	Take the GRE	Apply	Site Visit	Accepting the Offer

STEP 1: IDENTIFYING PROGRAMS OF INTEREST

Throughout your many years as a graduate student, you will be working as a research assistant in a single lab. Additionally, graduate school applications are not only time consuming, but typically include an application fee ($0-$175). For these reasons, your first step in the application process should be to identify programs with research areas and lab groups of interest to you. There are many different approaches, and you should make use of as many as possible.

Internet Search

When just starting out, you can conduct a simple internet search from graduate program rankings. From there, you can browse program websites and begin to narrow down your selections. Each website will have a link to faculty bios and research group webpages. Here you can learn about the research being conducted in different labs and narrow down your graduate program search even further by including specific labs at specific programs.

Faculty Members

If graduate school has been on your radar throughout your undergraduate career, you have most likely completed a research project in one or more labs within your department. Reach out to your faculty advisor to discuss research groups outside of your institution and within your field of research. Faculty members may even be willing to reach out directly to a contact on your behalf.

Just like we established the importance of a robust network for landing a job in industry, faculty are your best network for landing a graduate school position. Once you have identified a few research labs that you are interested in, you should reach out to the faculty member through email to learn more about their research. Figure 5.1 shows a sample email. If you make a good impression to a faculty member, they could put in a good word with the selection committee. Be sure to send a follow-up email after your application has been submitted.

Dear Dr. Ravits,

I am a senior chemical engineering student at Tulane University and have recently begun looking into graduate programs. The program at UT Austin is on the top of my list, particularly because of the research that your group is conducting. During my time at Tulane, I have worked as an undergraduate researcher in the lab of Dr. Julie Albert which has sparked my desire to pursue graduate studies within the field of polymers. I am particularly drawn to your lab's work with semi-crystalline morphologies and your recent publication in *Polymer Science*.

    If you have time for a brief meeting, I would love to learn more about your lab and opportunities available within your research group for incoming graduate students.

Thank you, and I look forward to hearing from you.

Katherine A. Smith
Tulane University, Class of 2022
ksmith@tulane.edu

Figure 5.1. Example of Request for Graduate Research

STEP 2: TAKE THE GRE

The GRE or Graduate Record Examination (https://www.ets.org/gre) is a standardized test required for admission to many graduate programs. There are numerous GRE prep courses, study guides, and practice tests available online, and you should begin prepping during the spring of your junior year. Plan to take the exam during the summer before senior year so you have plenty of time to retake the exam if you are not satisfied with your score.

STEP 3: APPLY

When applying for graduate school, your main focus will be demonstrating that you have the potential to conduct independent research. Most graduate programs will require the following documents with your application:

1. **Curriculum vitae (CV)**
2. **Personal statement**
3. **Letters of recommendation**
4. **GRE scores**
5. **Official transcript**

Most applications are due by January of your senior year, so you should plan accordingly to ensure all documents are submitted before the deadline. To keep organized, you should make a spreadsheet with relevant information for each program—including deadlines and required documentation. You will need to request letters of recommendation, an official transcript from your undergraduate institution, and GRE scores before the application due date.

STEP 4: OFFERS AND SITE VISITS

Now that all applications are submitted, the next step is to wait patiently for decisions from each program. For programs which you have been accepted to, you will receive an offer letter and, most likely, an invitation to visit the campus. For students with multiple acceptances, the spring semester of senior year can be busy but rewarding as you have the opportunity to see first-hand the research labs of interest to you. The trips are fully funded and a fantastic opportunity to find the ideal fit for you.

STEP 5: ACCEPTING THE OFFER

Your acceptance letter will likely detail the steps to accept the offer and will not require a formal acceptance until after the site visit. See Chapter 3.1 for tips on selecting between multiple offers.

CHAPTER 3.6

GUIDELINES FOR GRADUATE SCHOOL APPLICATIONS DOCUMENTS

GUIDE TO CURRICULUM VITAE (CV)

You are most likely more familiar with the term resume than CV, but the two have many similarities. A CV is basically a resume designed for academia and contains many of the same components. Just like the goal of a resume was to sell yourself for a position in industry, the goal of the CV is to sell yourself for a graduate school position. Therefore, the CV will focus on your academic and research history, as well as any additional relevant accomplishments. Another major difference between resumes and CVs is length. You learned in Chapter 3.2 that early-career resumes should not exceed one page, while a CV can be any length (typically one to two pages for an early-career researcher like you).

Formatting

Page Layout

You have many options for the overall layout of your CV. Be sure that the layout you choose is easy-to-read and draws the reader's eye to the most important information. For readability, you want your section headings left-aligned with your section information indented. The level of indentation is up to you, but be sure the finished product is pleasing to the eye. This means plenty of white space and varied spacing, fonts, and formatting. Again, use tables to efficiently align your info as demonstrated in Figure 3.2.1. Additional formatting considerations include:

- **Spacing:** Single-spaced
- **Font:** Professional, no less than 11 point
- **Section Titles:** Informative and attention grabbing
- **Page numbers:** Unlike resumes, CVs should include page numbers. These can be right-aligned in the header or right- or center-aligned in the footer.

CV Sections

Like a resume, you will divide your CV into relevant sections arranged in the standard order outlined below:

- Heading
- Education
- Experience
- Personalized Sections

Formatting and content information for the above sections are detailed in Chapter 3.2, so this section will elaborate on additional sections relevant to a research-focused career.

Research or Teaching Experience

Since your CV focuses on your preparation for academia, your Experience section may look different than that of a student applying for jobs in industry. You will most likely have undergraduate experience working in a lab, and you may have even participated in a STEM outreach workshop or event. Depending on the type and amount of information that you have to present, you may divide your Experience into more than one section. For example, Work Experience to cover general employment and Research Experience for academic experience. Again, be sure to include bullet points to demonstrate major accomplishments and skills obtained.

Publications

Students with significant research experience may have had the opportunity to author or co-author a journal article within their research group. You should list these in their own Publications section using the preferred citation style in your field.

Fellowships & Grants

Include any research grants and fellowships that you have received. If you are unsure how to begin applying for a research grant, the best place is within your institution. A simple internet search will direct you to your institution's information on funding sources for undergraduates.

PERSONAL STATEMENT

Another important component in your graduate school application is your personal statement. While your CV succinctly presents your experience and qualifications for an academic career, your aptly named personal statement is an opportunity to get more personal. Here you will share with the admissions committee your academic goals for graduate school and your reasons for pursuing a graduate degree.

Each school will have specific guidelines for what the admissions committee is looking for in your personal statement, but they typically fall into two categories: 1) an open-ended personal statement allowing you freedom on what to write, or 2) a more directed response to specific questions. Therefore, you will need to tailor your statement to each school you apply to—depending on requirements.

To prepare for writing your personal statement, consider your past experiences and how they relate to your career goals. Avoid simply rehashing your CV and be sure to further explain how each of your accomplishments better prepares you for graduate study. Do not forget to customize your information to each program that you are applying for. Include details when discussing any projects that you have worked on and the tangible results of that work. This is also an opportunity to callout faculty members within your desired program and identify how your interests align.

Katherine A. Smith

300 Lindy Boggs Building, New Orleans, LA 70118
(504) 865-8000 | ksmith@tulane.edu | www.linkedin.com/in/KASmithTulane

Education

Bachelor of Science in Chemical Engineering, Minor in Mathematics May 2022
Tulane University, New Orleans, LA GPA 3.87

Honors: Shell Undergraduate Award • AIChE Activity Award • Tulane Distinguished Scholars Award

Research Experience

Undergraduate Research Assistant August 2019 – Present
Albert Laboratory, Tulane University, New Orleans, LA
- Related the acid-driven degradation rate of polycaprolactone thin films to their semi-crystalline morphology
- Analyzed morphology and degree of crystallinity using optical microscopy, atomic force microscopy, and grazing-incidence X-ray scattering
- Presented research at the American Physical Society March Meeting DPOLY Poster Session
- Won first place undergraduate poster award at Tulane University School of Science and Engineering poster session

Research and Development Intern Summer 2020
ReactWell, New Orleans, LA
- Identified and addressed production failures using literature review, design of experiments, and data analysis
- Developed an optimum mixture of additives designed to increase energy storage in lithium-ion batteries

Publications

FG Miller, **KA Smith**, TR Green, JH Gilbert, BA Brown. Optimization of lithium-ion battery additives. J Mat Sci, 10:231 (2022).

Grants and Fellowships

Center for Engaged Learning and Teaching Undergraduate Research Grant, Tulane University 2021

Leadership and Campus Involvement

Tau Beta Pi Honors Engineering Society Spring 2019 – Present
Corresponding Secretary (Spring 2021 – Present)

American Institute of Chemical Engineers Spring 2019 – Present
President (Spring 2021 – Present), Treasurer (Spring 2020 – Spring 2021)
Class Representative (Fall 2019 – Spring 2020)

Theta Tau Professional Engineering Fraternity Fall 2019 – Present
Chemical Engineering Chair (Spring 2021 – Present)

Tulane Engineers without Borders Spring 2020 – Present

Boys at Tulane in STEM May 2021
Workshop Leader (May 2021)

Figure 6.1. Example of an early-career CV

Formatting considerations

Individual programs will provide guidelines regarding length requirements, but personal statements are typically short (one page). Use a professional font, and call attention to the most important items using formatting such as bolding and underlining.

LETTERS OF RECOMMENDATION

A graduate school application packet will always require letters of recommendation. Your choice of whom to ask to write a letter will be important, so they should be selected carefully. You will typically need to select three letter writers, and they should be faculty members who know you well and can attest to your strong character and ability to conduct research in a graduate program. Faculty members that you have worked directly with in a research setting should be your first go-to, followed by faculty members whom you have a strong educational relationship with.

Faculty members are often writing letters for multiple students to multiple universities, so you should make it as easy as possible for them. When asking a faculty member to write a letter, always provide advanced notice (at least two weeks), and include a copy of your most recent CV. If you are applying to numerous schools, include a list of each school and due date. Most importantly, do not forget to send a thank you note.

CHAPTER 3.7

GUIDE TO LINKEDIN

The goal of a LinkedIn profile

After polishing your resume and cover letter, you might be wondering why you still need to add a LinkedIn profile to your application arsenal. In Chapter 3.1, you learned the importance of establishing a vibrant professional network. A strong LinkedIn profile will help you expand your network, as well as give you a leg-up in the job hunt category through its robust job board and search functions. This section will provide guidelines for designing your LinkedIn profile, as well as tips and tricks for making the most of the platform.

Getting started with your profile

After signing up for a free LinkedIn profile, you can begin personalizing your profile page. LinkedIn has made this easy for you with a series of guided steps to complete the main items of your profile. Figure 7.1 shows a new profile, and each item is discussed in more detail in this chapter.

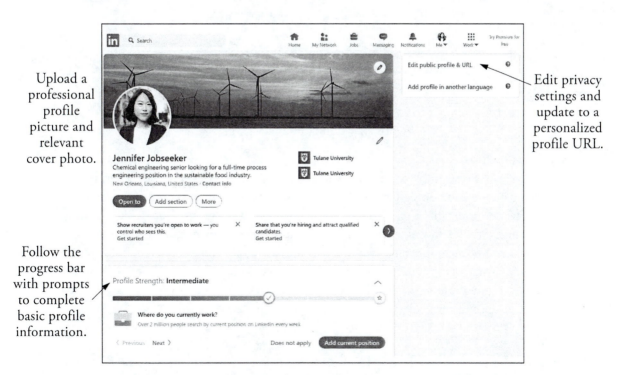

Upload a professional profile picture and relevant cover photo.

Edit privacy settings and update to a personalized profile URL.

Follow the progress bar with prompts to complete basic profile information.

Figure 7.1 Basic profile template after signing up for a LinkedIn account

273

Profile Sections

Chapter 3.2 outlined the standard sections for resumes. Similarly, LinkedIn makes it easy to include these same sections in your public profile. You can select which sections to display on your profile and edit each as appropriate. To begin, you should add each of the sections that you included within your resume and expand from there. The Add Profile Section button (Fig. 7.2) provides all additional options available to you.

Edit section information by selecting the pencil icon.

Customize which profile sections you want to be viewable on your profile.

Figure 7.2. Adding Profile Sections

Building Sections

Once you create a new profile section, you can add individual sections within the main section by clicking on the addition sign, shown in Figure 7.3. LinkedIn makes it easy to incorporate your relevant information by guiding you through a series of questions. Once all information has been input, LinkedIn will create and format the section automatically. You should input your information (including bullet point items) as they appear on your resume. Figure 7.3 demonstrates what an Experience item within your LinkedIn profile would look like if pulling information from a resume section. From there, you can expand upon what is presented by including media and any additional items not included in your resume.

Profile Photo

Your LinkedIn profile is a professional representation of yourself. Therefore, your profile photo should look the part. You do not need a professional photographer to generate a professional

Resume Section

<u>**Experience**</u>

Research Assistant   June 2018 – Present

Albert Research Group – Tulane University | New Orleans, LA

- Wrote Python code to determine the Order-Disorder Transition (ODT) points for several polymer blends
- Calculated radial distribution functions (RDFs) and analyzed output data to infer system characteristics, such as spacing of lamellar domains

LinkedIn Section

Experience   +

Research Assistant   ✎

Tulane University · Part-time

Jun 2020 – Present · 9 mos

New Orleans, Louisiana, United States

- Conducted independent research in the lab of Dr. Julie Albert within the Chemical and Biomolecular Engineering Department.
- Wrote Python code to determine the Order-Disorder-Transition (ODT) points for several polymer blends.
- Calculated radial distribution functions (RDFs) and analyzed output data to infer system characteristics, such as spacing of lamellar domains.

Figure 7.3 Layout from a resume to a LinkedIn profile

looking headshot. Simply have a friend or family member take a few photos of you. Choose a simple background and professional attire.

Background Photo

The background photo is an opportunity to give your profile some personality. Again, the selected photo should be professional and tailored to your interest. Perhaps something related to engineering or science, or your location. LinkedIn recommends using png, jpg, or gifs with resolution of 1400x425.

Headline

The goal of a successful headline is to grab your audience, and it is how people will find you when searching by specific keywords. Try to demonstrate why you are someone that people want to connect with. Start by highlighting your major, current position, and university or company, then expand by highlighting your skills.

Summary (About)

Like a cover letter to a resume, the About section provides you an opportunity to focus on a few select items to really sell yourself. Use the space to expand on what you have written in your headline. Make it professional, easy-to-read, and give the reader a taste of your personality.

About   ✎

During my studies at Tulane University, I have been fascinated--both inside and outside of the classroom--in the emerging technology of alternative proteins. As a chemical engineering major, I have developed a strong set of problem-solving and technical skills, and a knack for identifying inefficiencies and process optimization.

With the goal of connecting these skills with my passion for sustainability, I've tailored my coursework and work experiences to prepare myself for a role in the sustainable food industry. I'm looking forward to putting my experience to good use in the next chapter of my career by working in an industry that has the potential to profoundly change our world for the better.

Figure 7.4 A sample About section

Media

The Media option in LinkedIn allows you to add or link to external documents, photos, websites, videos, and presentations. Adding media to your profile is an effective way to provide tangible examples of your accomplishments. In some cases, you can embed media directly into a section. For example, within your Experience section you could include an image of a research poster presented while working as an undergraduate research assistant. You could also upload directly to LinkedIn's Featured section designed to let you showcase posts, documents, media, and website. Additional ideas for media include figures from school projects, publications, links to press releases, photos of physical projects, and class or work presentations.

Additional Sections

Just like with your resume, once you have covered your education and experience, you have flexibility in which additional information to include. Browse the Add Profile Section list for ideas, and remember to include any information that highlights your accomplishments.

Using the Search Function to Network

The search function within LinkedIn is a powerful tool to help you find contacts in your area and beyond. Use the People You May Know section to identify contacts of contacts and branch out from people in your immediate network by using LinkedIn's robust search function. The advanced search function lets you locate potential contacts by job, location, school, etc. and is the best way to find alumni from your institution. Figure 7.5 demonstrates how to search for People with affiliations at both Tesla and Tulane University. Once you have identified a potential new contact, send a connect request. Be sure to personalize your message by letting the contact know where you met, similar interests, etc. Figure 4.1 provides an example of a networking message to a first-time contact.

Begin your search with a company of interest.

Select *People* to limit the search to potential contacts, and *All Filters* to open additional search options.

Narrow your search further by selecting additional filters. Specify School if you are looking to identify alumni from your university.

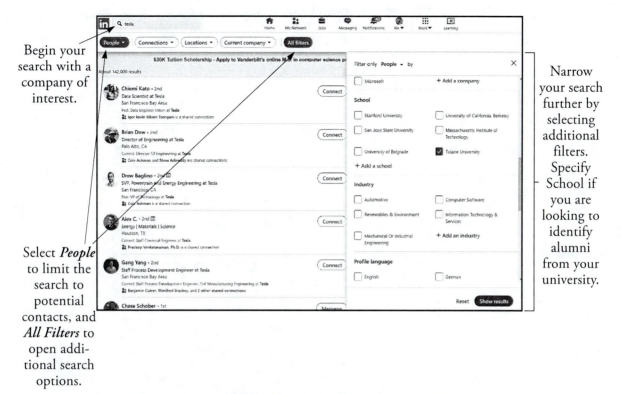

Figure 7.5 Sample LinkedIn search to identify potential contacts

A NOTE ABOUT SOCIAL MEDIA

In the age of social media, you have most likely heard a story or two about someone losing their job due to an inappropriate or unprofessional post. With that in mind, be conscious of your online and social media presence. Before you begin the job hunt, review your privacy settings on all social media accounts to keep your personal life private. Conduct a Google search to see what shows up in addition to your professional LinkedIn profile--potential employers will often do the same.

CHAPTER 3.8

ADDITIONAL RESOURCES

There are numerous online resources for students interested in learning more about the topics covered in this book and beyond.

Purdue Online Writing Lab (OWL) offers extensive resources in all topics related to writing, including Job Search and Professional Writing.
Website: https://owl.purdue.edu/owl/purdue_owl.html

LinkedIn Learning provides access to a vast library of on-demand content from industry leaders within business and technology. The platform requires a monthly subscription, but many universities provide free access to students. Simply check your career center's website for information on how to access.
Website: www.LinkedIn.com/learning/

Coursera is a free online provider of courses from top universities and companies. Search for courses in professional development or even consider learning a new skill for your resume.
Website: https://www.coursera.org/

Part Four:
Writing on the Job

CHAPTER 4.1

GUIDE TO REPORTS IN THE WORKPLACE

Introduction

Reports that are prepared for workplace projects superficially resemble the reports that students prepare for undergraduate projects. That is, they use the same kinds of section headings, they present information in roughly the same order, and they use the same kinds of plots and diagrams to display results. Workplace reports differ from student reports because they are more complex, and this complexity can involve both the subject matter and the audience.

Subject matter. Undergraduate laboratory projects almost always present students with well-defined goals and procedures, and workplace projects do not always do this. With the exception of form-driven quality tests, workplace projects will likely require you both to determine your project's technical goals and to define how particular tasks help you meet those goals. You will usually present these definitions in your report's Introduction section, which consequently will tend to be longer than the Introductions of your undergraduate reports.

Audience. Undergraduate reports are written for your instructors. These are people who you know and who understand your projects even before you submit your reports. Workplace reports, in contrast, may be read by people other than your supervisor. You may not know these people, and you cannot assume they are experts in your field. Yet you should make sure your reports are accessible to such outside readers because they can have a significant impact on your work—they may include government officials who must approve your work, corporate lawyers who might defend your work, or finance officials who control your project's funding.

You can best solve the problems of audience and subject matter by writing reports that are complete and by using images that provide context. Completeness in workplace reports is managed at the beginning of the document. Here your Introduction should summarize the task you received from your client or supervisor, and it should explain how your project steps help you meet the objectives that have been assigned to you.

Presenting Images:

Many of your readers will be unfamiliar with your project. They will have trouble visualizing both the things you have made and the environment where these things will be used. You can help these readers understand your work by displaying images, such as photographs, drawings and diagrams. As you design images for the different sections of your reports, you should assume that your figures do different jobs for the reader and provide different kinds of information at different points in your report. In an Introduction, for example, your job is to name your topic and orient your reader, so figures should display with few details the subject of your work or the object you were asked to modify. A Result or Design section should provide substantial information about your work, so figures should display, for example, detailed views of the device you have designed or examined. In the same way, an Analysis section might describe safety calculations, so it might display a mathematical simplification of the system you have developed or tested.

To assure that non-expert readers can understand your work, it is useful to provide context information for all sections of your reports. This often requires that you use pairs of illustrations, where a diagram of new, technical information is displayed beside a photograph or drawing of the device or system as it appears in common use. The use of paired illustrations greatly simplifies the job of the non-expert reader in a technical report. You can further simplify that reader's task by preparing figure descriptions that define what is displayed in your paired figures.

In addition to selecting figures that present an appropriate amount of information, it is important to label the components in those figures and describe them in text. Those text descriptions should be brief, and they should explicitly call out and explain the labeled parts of the drawings or photographs.

Below we present three sets of paired figures from a sample report that will be presented in a later chapter. Using these figures and their text descriptions, we will explain how you can easily and clearly present complex work to readers who are not professionals in your field.

An Introductory Figure.
The first job of a report's Introduction section is to name the subject of study and define the problem that the project addresses. Because readers are not always technical experts, that definition needs to be concrete, and it ought to connect the technical task to an object or concept that most readers should have seen or heard of before. Figure 1 shows the opening page of our example report on the redesign of a modular bridge expansion joint. Here, four sentences and one figure work together to present all of this information.

1a) Name the subject for non-experts.
The first sentence defines Modular Bridge Expansion Joints in functional terms, and it cites Figure 1, which shows on the left a typical expansion joint on a highway bridge.

1b) Identify the subject for experts.
The second sentence defines the expansion joint components that are important in this project. These components are labeled in the diagram on the right side of the figure, and these labels correspond to the terms used in the text. That diagram is aligned with the photograph in such a way that the roadbed, joint and concrete barriers have roughly the same arrangement in each image. This alignment makes it easy for all readers to identify the way the diagram maps onto the photograph of a typical joint.

1c) Define the problem.

The third and fourth sentences define the problem, with sentence three describing the materials and fastening methods and sentence four naming the specific problem of fatigue under cyclic loading, as represented by the trailer shown in the right-side photograph.

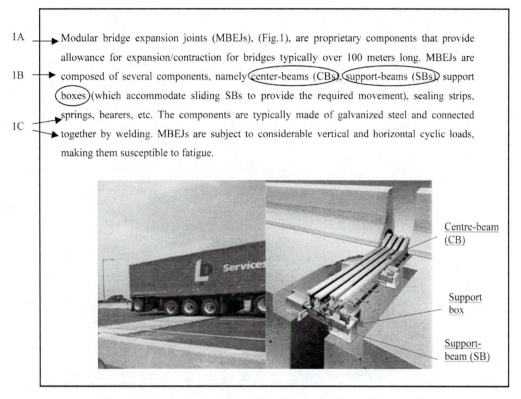

Figure 1. Introductory figure with text description.

A design or results figure

Results, like introductory information, are best presented using a combination of diagrams and words. Figure 2, the design of our sample report, offers an excellent example of such a presentation, as it is complete, direct, and short. The diagram is lightly dimensioned, as is reasonable for an initial discussion of a design concept. The diagram is fully labeled; the four components of the proposed design are labeled and called out in the text description. That description fills three sentences. The first cites the figure and defines its topic. The next two sentences list and explain the labeled components of the diagram; the circled terms call out labels in the drawing. Because an existing device is being modified, the explanations focus on methods and points of attachment. Specifically, the discussion of the FRP box section states that it is to be "*epoxied* to the *bottom flange* of the steel beam."

Figure 6 shows the proposed splice for a single span centre-beam. An FRP box section would be epoxied to the bottom flange of the steel beam. The web of the centre-beam would be spliced using steel plates and pretensioned steel bolts to create a slip-critical connection.

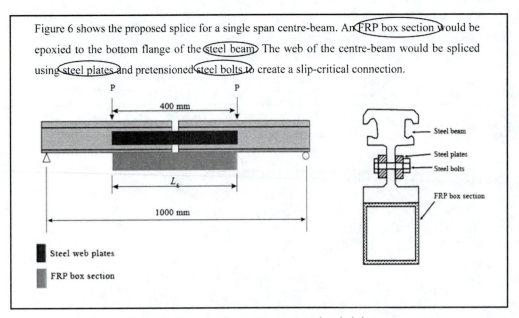

Figure 2. Figure and text presenting a detailed design.

The current designs for centre-beam field splices, whether of the hinge- type (Fig. 3) or moment-resisting type (Fig. 4), do not possess the same fatigue life as that of the other components of the MBEJ and thus cause premature, costly failures in the installed MBEJs.

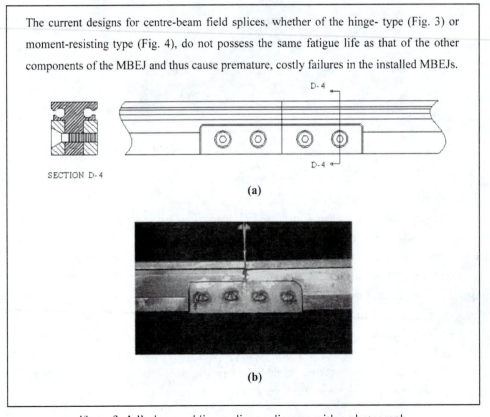

Figure 3. A Background figure aligns a diagram with a photograph.

A background figure

A review of the prior art with regard to a problem can help to clarify a problem and to explain the advantages of the solution that your team might propose. In Figure 3, the prior art is presented visually, with beam splices represented both photographically and diagrammatically. The figure description here is very short, but the photograph and the diagram have been prepared such that the splice and the bolts very prominently display the points of alignment, while the text identifies fatigue as a problem in this form of splice.

Tables

To make tabular information accessible for the non-expert, it is best to keep tables tightly focused and to describe them thoroughly. Figure 4 presents the results of a simple analysis, and it demonstrates how to do both of these things. In this case, a calculation was performed to determine the length of a splice that would be attached to reinforce a support beam, and the results were presented in tabular form. The table contains two kinds of information: the deflection of an intact beam and the calculated deflection of a spliced beam. The goal would be to determine the size of a splice so that the beam-splice system deflects no more than an

---

The model was first used to determine the length of FRP required to ensure the stiffness of the spliced beam is similar to that of an intact beam. In this case, a GFRP beam with $E_{11} = 120$ GPa was assumed. Table 4 shows the results.

Table 4. Comparative deflection analysis for beam splice sizes

Beam				δ_{max} (mm)
Intact beam				1.515
Spliced beam (using GFRP tube)		$E = 120$ GPa	$L_s = 200$ mm	1.982
			$L_s = 300$ mm	1.741
			$L_s = 400$ mm	1.484

The maximum deflection of the intact beam under the truck loading is 1.515 mm. The results indicate that the GFRP box beam ($E = 120$ GPa) must be at least 400 mm long to ensure that the maximum deflection of the spliced beam is similar to that of an intact beam. The typical span of a centre-beam is 1 metre, so a 400 mm FRP beam would still have 300 mm clearance on each side for installation in the field.

Figure 4. An example results table with text description.

intact, unmodified beam. The table compares the deflection of an unmodified, intact beam with the deflections attributed to beams having three different sizes of splice. This table justifies a decision to use a splice length of 400 mm because this splice size yields a deflection close—to but less than—the deflection of an unspliced beam.

This table is small, and it is efficient. It presents an analytical result that justifies a decision to set the splice size at 400 mm. It does this by offering two kinds of comparative information—the benchmark value of 1.515 mm deflection for an unmodified, intact beam, and the comparison deflections for splices of 200 and 300 mm.

The text description of the table is complete, but the description is brief because each sentence performs a specific task. The first presents the goal of the calculation, and the second presents the assumptions that were used. The second paragraph, below the table display, speaks directly to the contents of the table, calling out the benchmark value of the intact beam, the assumed value of E11 and the selection of 400 mm as the best size for the splice. The analytical result is captured in a table of five rows, while the text fully speaks to the table in the space of seven lines.

CHAPTER 4.2

GUIDE TO REPORT ORGANIZATION

Workplace reports are organized and assembled according to the same format scheme as is used in undergraduate student reports. Workplace reports differ from undergraduate reports due to matters of scale rather than of kind. Accordingly, the sample report presented here is subdivided into roughly the same headed sections as are found in student reports:

Introduction
Methods and Organization
Results
Analysis
Conclusion

Below we describe briefly what information you should present in each of the main headed sections of a typical report. Following this we present a sample project report that we have annotated to call attention to the way information is presented in each of these sections.

Introduction

The Introduction to a report should explain what work has been performed and why it has been performed. In most cases, report introductions are built up from five pieces of information

1. The name of the Topic
2. The Motivation or Need for the work,
3. Related Background and Problem,
4. The Goal of your work
5. The Scope of the report

This information should be as brief as is reasonably possible. To present this information, it is often helpful to compile your Introduction by recording your answers to these questions:

1. What did you work on?
2. Why was it necessary to do this work?
3. Has other work been done in this area?
4. What did you attempt to accomplish?
5. How much of your work is presented here?

This list reflects the questions that readers ask most often when they begin to read technical reports, and it presents the most common way to order these statements. However, small variations are common and reasonable. For example, when your report describes a completed project, the Goal and Scope of the report will probably be the same. In the same way, the Motivation and Background statements will be the same when you address a problem that others have tried and failed to solve. In our sample design report, these questions are addressed over the course of several pages, including both the **Introduction** section and the **Requirements for Fatigue-Resistant MBEJ Splice** section.

Workplace projects are commonly undertaken at the request of a client or a supervisor. When this is the case, your Introduction should first thoroughly review the specifics of that request. After you complete that review, you should add any other information that is pertinent to the task, such as attempts that others have made to solve the problem. When you present your own background research as well as your client's description of a problem area, you should present the client's problem description first, and you should make it clear where the client's problem description concludes and your input begins. A simple way to draw this distinction is to place your contribution in a separate paragraph, and to open that paragraph with a transitional statement such as this: "In order to address these concerns, it is important to consider the following additional factors…"

Project Formulation/Goals

Your last job in an Introduction is to identify the particular goals of your work and the particular challenges you addressed in order to meet those goals. This is commonly offered in the form of a list that describes your project work and that corresponds to the list of sections in the rest of your report. In this report, the section **Requirements for Fatigue-Resistant MBEJ Splice** presents a concise formulation of the project's goals.

Workplace reports do not always provide separate **Methods** sections because workplace projects do not always involve experimentation. In design and analysis reports, such as that presented here, methods information is distributed through the task and analysis sections of the document, while only a general overview of your project's organization is presented at the point of the task formulation. In this example report, that organizational information is compressed to two sentences at the end of the section **Requirements for Fatigue-Resistant MBEJ Splice**.

Design Description or Results Description.

Because projects can take many forms, the findings or results presentations can have many flavors. On some projects, you present objects or facilities that you have examined, while for other projects you might use plots and data tables to display your accomplishments. In either case, your results are usually best presented visually, and your tasks in the report are to display those results effectively and to explain them so that readers can see what is significant about them.

The standards for preparing displays are described elsewhere in this book. Here it is sufficient to remind you that figures need to be sized so that they are large enough to be clearly visible in printed form. It is also a good idea to highlight important areas on your images by imposing one or two arrows or circles. When you do this, you need to explicitly speak to those highlights in your text.

Any time you display a figure, you should provide a descriptive caption, which answers the question "What is this?" For professional reports you should also provide a text description of the display that explicitly defines what is in the display and what readers are expected to derive from it. To do this efficiently, you should describe figures according to this simple checklist.

1. Cite the figure by number in your text
2. State the figure's purpose
3. List the labeled components of the figure
4. Discuss the figure:
   a. Explain what the image demonstrates (for plots or tables)
   or
   b. Explain how the device operates (for objects or devices)

Other types of discussion are acceptable, of course, depending on the image. Our sample report offers an excellent design presentation in the headed section **Proposed FRP Splice Design**.

Analysis or Validation

Results presentations usually are supported by some type of discussion or analysis that either explains the results that have been obtained or that validates a design for feasibility or safety. Analysis statements can be short or long, depending on the complexity of the process. Analysis sections should answer simple questions about your work. These questions may include:

1. How do you know your experimental results are correct?
2. How do you know your design will be safe?
3. How do your results compare with benchmarks in your area?

Each of these questions speaks to a different type of project, but they all speak to the need for any technical professional to evaluate results dispassionately and thoroughly.

An analysis presentation can resemble a short experimental presentation, for it opens with a task definition, and it must account for the methods, including assumptions, models and values that were used. It should also indicate areas where the problem has been adjusted for simplicity, where values have been selected in order to be conservative, or the like.

Our sample design report presents an extensive analysis in the section **Numerical Investigation of Proposed Splice Detail**.

Closing Summary

A workplace report should close with two small chunks of information: a point summary and a list of action items.

Summary. A closing section should first summarize the significant findings that have been mentioned elsewhere in the report. The task of this summary is simply to collect in one location those points that readers need to remember and that they may need to present to their superiors. These summary statements should be short; you should limit these to two lines of text per point, and you should consider formatting them as bullet points.

Action items. The second part of a closing section should specifically list any actions that you recommend that the client take, either to address problems from your point list or to schedule follow-up reviews at a later time.

Because our sample report fulfills a contract with an outside agency, there are no action items for our author to present. Consequently, our example closes with a summary only.

CHAPTER 4.3

EXAMPLE OF A DESIGN REPORT FOR A WORKPLACE PROJECT

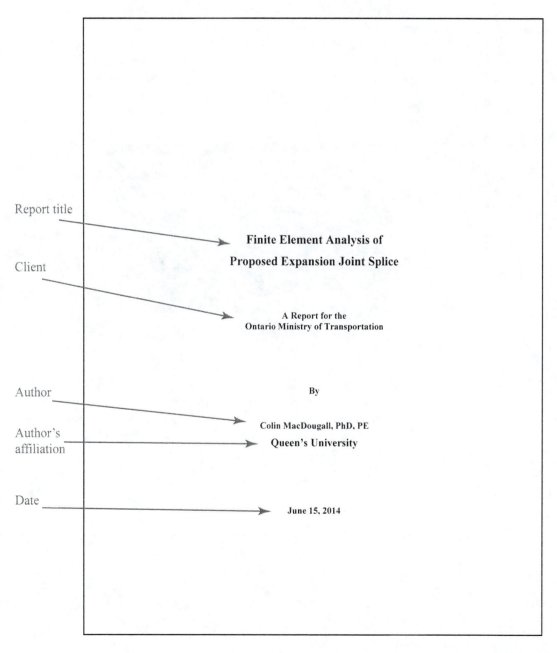

Report title → **Finite Element Analysis of Proposed Expansion Joint Splice**

Client → A Report for the Ontario Ministry of Transportation

Author → By Colin MacDougall, PhD, PE

Author's affiliation → Queen's University

Date → June 15, 2014

The topic is defined in terms that non-expert readers can understand before professional terminology is introduced

Introduction

Modular bridge expansion joints (MBEJs), as in Fig.1, are proprietary components that provide allowance for expansion and contraction for bridges typically over 100 meters long. MBEJs are composed of several components, namely center-beams (CBs), support-beams (SBs), support boxes (which accommodate sliding SBs to provide the required movement), sealing strips, springs, and bearings. The components are typically made of galvanized steel and are connected together by welding. MBEJs are subject to considerable vertical and horizontal cyclic loads, making them susceptible to fatigue.

Centre-beam (CB)

Support box

Support-beam (SB)

Figure 1. Modular Bridge expansion joint (right picture adopted from [1])

The specific problem is identified

Premature failures due to fatigue have been reported in installed MBEJs [2 & 3]. These failures require either repair or replacement. Exhaustive research by Dexter et al. [4] examined the behaviour of MBEJs under static and cyclic loading. Among many issues needing more research, as stated in NCHRP report 467, is a fatigue-resistant field splice for centre-beams. To minimize traffic blockage, the replacement of failed MBEJs needs

Related background research is presented

1

Background
continues
here

Other splice
methods are
evaluated

The project
issue is defined
as poor fatigue
life of splices

Goal is defined

Project scope:

* Requirements
* Design
  performance
  analysis

to be performed in several stages, during which one or two lanes may be closed. The installation of an MBEJ will typically be staged, and therefore these joints need to be spliced in the field. It is not always possible to have optimum quality control over the splice installation due to the limited working space, as in Fig. 2, and to time constraints. The current designs for centre-beam field splices, whether of the hinge-type, as in Fig. 3, or the moment-resisting type, as in Fig. 4, do not possess the same fatigue life as do the other components of the MBEJ and thus cause premature, costly failures in the installed MBEJs. Chaallal et al., thus, recommend that "an improved splice be designed" (2002) [5].

Figure 2. View within a modular bridge expansion joint.

The Ministry of Transportation of Ontario (MTO) has requested Queen's university researchers to examine this problem. The study is to involve industry partners in Ontario and Quebec. This report presents a new design for a center-beam field splice. The next section presents the requirements for the new design. The splice design is then briefly presented, after which a numerical investigation is described in order to validate the performance of the proposed splice. The results are summarized at the end of this report.

2

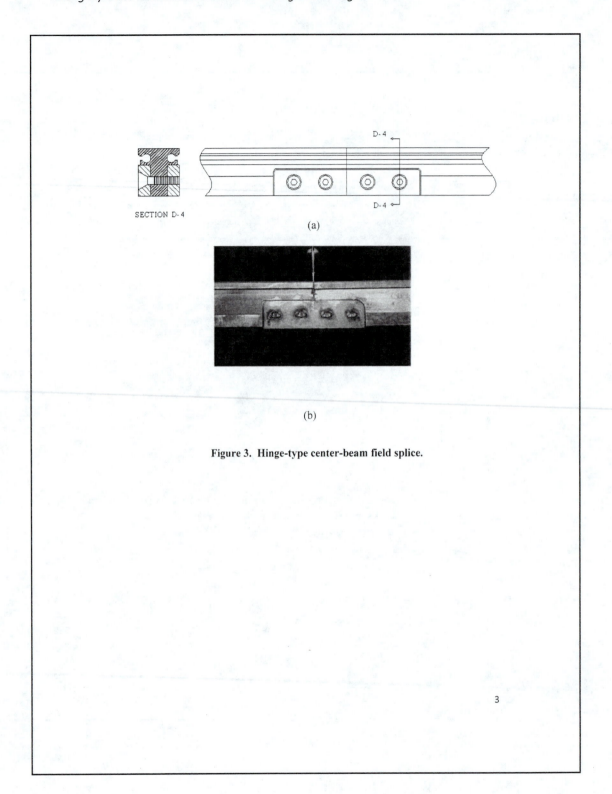

(a)

(b)

Figure 3. Hinge-type center-beam field splice.

3

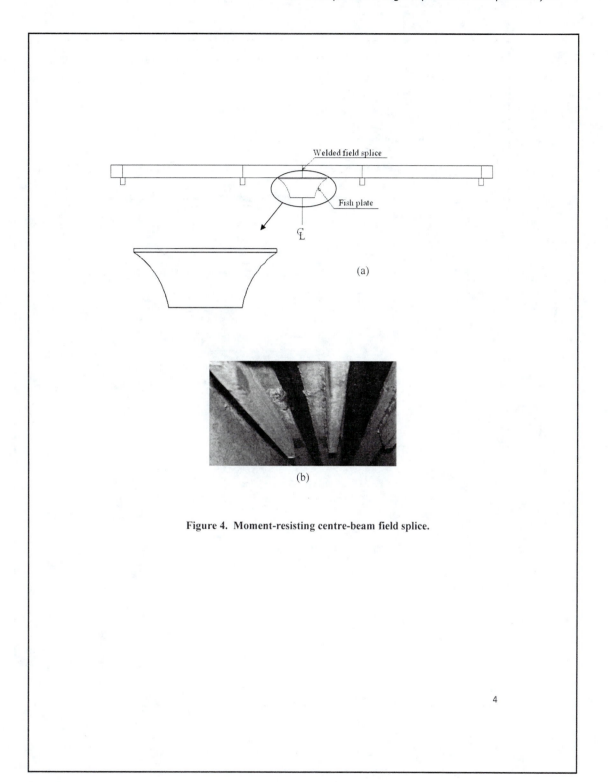

Figure 4. Moment-resisting centre-beam field splice.

Qualitative
design criteria.
Specifications
can be listed
in this area
as well

Requirements for Fatigue-Resistant MBEJ Splice

A new fatigue resistant design for an MBEJ should meet the following criteria:

- The detail should be versatile enough to accommodate the variety of proprietary centre-beams on the market.

- The detail must be feasible and easy to install in the field.

- In order to be consistent with the fatigue resistance of the rest of the MBEJ system, the splice detail should be at least Category C [5].

- The detail must be resistant to all dynamic and static effects of both vertical and horizontal loads and must be durable over the design life of the MBEJ.

The design presented below is intended to be versatile and easy to install in the field. Finite element model predictions presented in the second half of this paper demonstrate that it is resistant to fatigue and to loading effects.

5

Design
description
with a
figure:

* Solution is
named and
motivated

* Figure is
cited and
identified

* Labeled
elements of
the figure
are called out

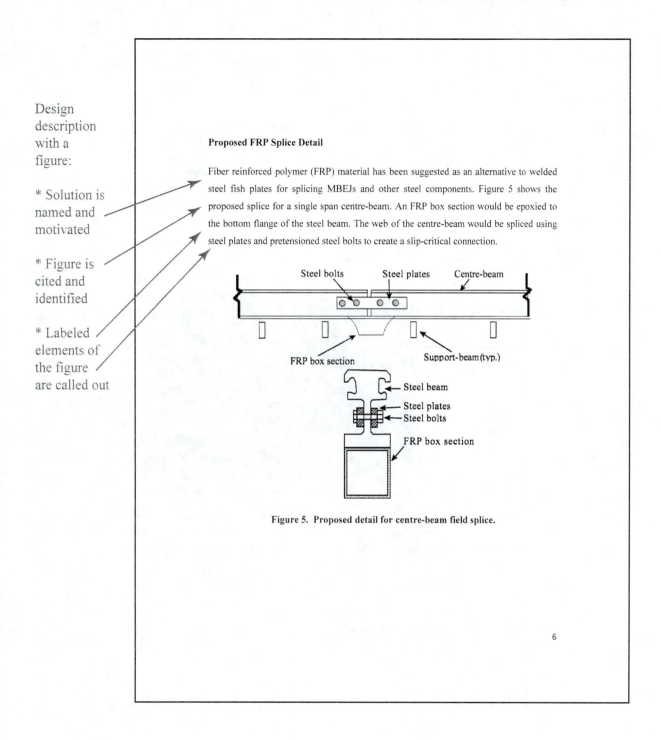

Proposed FRP Splice Detail

Fiber reinforced polymer (FRP) material has been suggested as an alternative to welded
steel fish plates for splicing MBEJs and other steel components. Figure 5 shows the
proposed splice for a single span centre-beam. An FRP box section would be epoxied to
the bottom flange of the steel beam. The web of the centre-beam would be spliced using
steel plates and pretensioned steel bolts to create a slip-critical connection.

Figure 5. Proposed detail for centre-beam field splice.

6

299

An analysis
presentation
must define:

1) Methods
and
assumptions,
2) Material
values,
3) Loads that
must be
accounted for

1) Methods and
assumptions
are defined
first

Finite Element Model Investigation of Proposed Splice Detail

In order to assess the feasibility of the proposed splice detail, a finite element model was developed and investigated. The finite element model was implemented in ABAQUS and is shown in Figure 6. Although the centre-beams of MBEJs are typically continuous over several spans, it was decided to model only a single simply-supported span to reduce the size and complexity of the model. In the future, the effects of continuity will be investigated. The span of the beam is 1 metre, which is typical of the centre-beams for an MBEJ, and pin and roller supports were assumed.

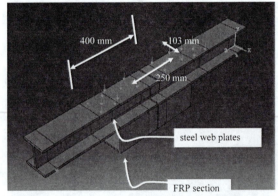

Figure 6. Spliced center-beam modelled using ABAQUS

In developing the finite element model, 10-node modified quadratic tetrahedron (C3D10M) elements have been used to mesh the 3D model. The steel was modeled as a homogeneous material having $E = 200$ GPa and the FRP was modeled as an orthotropic material using the data provide by the supplier. The steel beam dimensions were selected to be similar to those of typical MBEJ centre-beams on the market (W100 × 19).

The FRP section was selected to have the same breadth as that of the center-beam flange. Both carbon fibre and glass fibre reinforced polymer could be used for this application,

7

2) Material
properties and
dimenstions are
specified, as
are testing
conditions

although glass fibre reinforced polymer (GFRP) is cheaper and more widely available than CFRP.

Two different glass fibres (GFRP) with E_{11} = 17.1 GPa and E_{11} = 120 GPa and one carbon fibre (CFRP) with E_{11} = 400 GPa were investigated. The first GFRP, which has a very low modulus compared to steel, was selected because it is readily available; it was used for static tests of the detail. For the GFRP model, a wall thickness of 6 mm was assumed. For the CFRP model, a wall thickness of 2 mm was assumed.

The steel web splice plates were assumed to have cross-sectional dimensions of 6 mm × 60 mm. As per typical practice, there is a five-millimetre gap between the two halves of the steel beams.

The loading on the model is based on the Canadian Highway Bridge Design Code CL-625-Ont truck loading [6]. This loading is typically used to design MBEJs. Fig. 7 shows the CL-625-ONT truck loads and spacing. Half of the axle load (that is, one wheel load) was applied to the splice in the form of two distributed loads, each resembling one of the two wheels comprising each side of the axle. The CL-625-ONT wheel patch of 600 mm x 250 mm was modified so that in the model the wheel load was applied on two areas of 250 × 103 mm with a spacing between the centers of 400 mm. These loads are indicated in Fig. 6 as well.

8

3) Loads are described using diagrams of truck dimensions aligned with diagrams of wheel loads

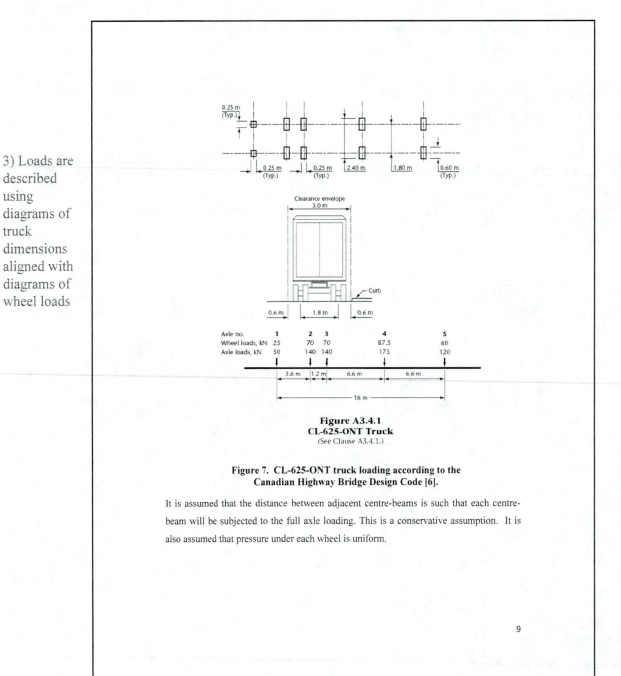

Figure A3.4.1
CL-625-ONT Truck
(See Clause A3.4.1.)

Figure 7. CL-625-ONT truck loading according to the
Canadian Highway Bridge Design Code [6].

It is assumed that the distance between adjacent centre-beams is such that each centre-beam will be subjected to the full axle loading. This is a conservative assumption. It is also assumed that pressure under each wheel is uniform.

9

302

3) Analytical
goals are
defined
to establish:

a) Stiffness
criteria, and

b) Fatigue
criteria

3a) Stiffness
analysis
begins

Result of
stiffness
analysis

3b) Fatigue
analysis begins
with
description of
Figure 8, the
analysis
output

Results

The main purpose of the splice is to provide a moment-resisting connection for the ends of the centre-beams. Theoretically, the spliced beam should have the same stiffness as an unspliced centre-beam. In addition, the stresses within the spliced beam should be low enough to ensure fatigue does not occur.

The maximum deflection at the midspan was used as the stiffness measurement. The model was first used to determine the length of FRP required to ensure that the stiffness of the spliced beam is similar to that of an intact beam. In this case, a GFRP beam with E_{11} = 120 GPa was assumed. Table 1 shows the results.

Table 1. The effect of FRP section length on the maximum deflection of the spliced beam.

Beam Condition	Length of FRP section (mm)	Maximum Deflection (mm)
Intact	-	1.515
Spliced	200	1.982
Spliced	300	1.741
Spliced	400	1.484

The maximum deflection of the intact beam under the truck loading is 1.515 mm. The results indicate that the GFRP box beam (E = 120 GPa) must be at least 400 mm long to ensure that the maximum deflection of the spliced beam is similar to that of an intact beam. The typical span of a centre-beam is 1 metre, so a 400 mm FRP beam would still have 300 mm clearance on each side for installation in the field.

Figure 8 shows the stress distribution of the spliced beam, in this case with an FRP box beam length of 400 mm. The spliced beam does not bend with uniform curvature, as the FRP box beam significantly stiffens the beam over the splice length. Regions of high stress are indentified in Figure 8: Location *A* is at the mid-span where the steel web splice plates meet the gap between the centre-beam ends; Location *B* is at the end tips of the

10

303

FRP section. The stresses at these locations are indicated in Table 2. The results in Table 2 are for 103 mm × 103 mm FRP box sections. Two different FRPs (glass FRP and carbon FRP) were considered, and the thickness t of each box section is indicated in the table.

Stress analysis results are presented in figures and tables. Table 2 is indexed to points A and B in Fibure 8

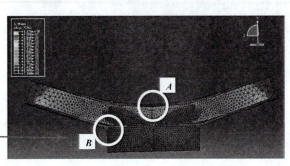

Figure 8. Stresses in spliced beam subjected to maximum truck loads.

Table 2. Maximum stresses in splice with a GFRP or CFRP section. Locations *A* and *B* indicated in Figure 8.

	σ_{max} (MPa)	
	GFRP (E=120 GPa; t=6 mm)	**CFRP** (E=400 GPa; t=2 mm)
A	140	168
B	167	327

Analysis results are compared to expected stresses

Steel will generally experience no fatigue for stress amplitudes below 50% of its static strength. Assuming 450 MPa for the tensile strength of the steel, the fatigue limit is 225

11

Modeling results are compared to calculated loads to validate the proposed splice

MPa. The maximum predicted stress in the steel, 168 MPa, occurs at Location *A* when a CFRP section is used. This stress is 33 % below the fatigue limit, indicating that fatigue in the steel will not occur.

The fatigue stress-life curve for GFRP composites, when plotted on log-log axes, is linear with a slope of 10%, and it does not exhibit a fatigue limit. An equivalent fatigue limit of 25% of the tensile strength is suggested for GFRP [6]. For GFRP with a tensile strength of 700 MPa, a fatigue strength of 175 MPa can be expected. A fatigue limit of 65% of the tensile strength is suggested for CFRP [6]. For CFRP with a tensile strength of 1431 MPa, a fatigue strength of 930 MPa can be expected. Comparison with the stresses at Location *B* in Table 2 indicates that the predicted stress in the GFRP is 5% below the fatigue limit and the predicted stress in the CFRP is 65% below the fatigue limit. Although the GFRP stresses are close to the fatigue limit, it should be noted that a factor of safety is already included in the fatigue limit suggested in [6]. Therefore, this would be an acceptable fatigue resistant design.

12

Concluding summary briefly restates the results that have been presented in the different sections of the report

Conclusions

An innovative method of splicing steel beams is investigated in this report. The splice detail uses bolted steel web plates and a fiber reinforced polymer box section epoxied to the beams. This detail was investigated for beams and loadings typically found in modular bridge expansion joints. Such a splice detail would be advantageous in this application because of its improved fatigue performance, which results from avoiding the use of welding.

A finite element model of the proposed splice detail was implemented and investigated. Glass fiber and carbon fiber were investigated as potential materials for the splice detail. The following conclusions result from the finite element analysis:

1) A minimum length of 400 mm for the fiber reinforced polymer section is needed to develop the required stiffness in the splice detail.

2) Glass fiber with a modulus of at least 120 GPa is needed to develop the required stiffness in the splice detail.

3) Both glass fiber and carbon fiber reinforced polymer can be used to develop a fatigue resistant splice detail.

13

References

[1] Hennegan and Associates. 2010. "Modular Expansion Joints." Electronic Product Brochure [online]. Obtained May 5, 2010. Available from henneganandassociates.com/expansion_joints.htm

[2] Roeder, C. W. (1998). "Fatigue and Dynamic Load Measurements on Modular Bridge Expansion Joints." *Construction and Building Materials.* 12(2-3), 143-150.

[3] Dexter, R. J., Osberg, C. B., and Mutziger, M. J. (2001). "Design, Specification, Installation, and Maintenance of Modular Bridge Expansion Joints." *Journal of Bridge Engineering.* 6(6), 529-538.

[4] NCHRP- Report 467. (2002). "Performance Testing for Modular Bridge Joint Systems." Edited by Dexter, R. J., Mutziger, M. J., and Osberg, C. B., National Cooperative Highway Research Program (NCHRP), Transportation Research Board (US), Washington, DC.

[5] Chaallal, O., Sieprawski, G., and Guizani, L. (2006). "Fatigue Performance of Modular Expansion Joints for Bridges." *Canadian Journal of Civil Engineering.* 33, 921-932.

[6] Can/CSA-S6-06. (2006). "Canadian Highway Bridge Design Code." Mississauga, ON; Cl. 3.8.3.2 and Annex A3.4.

14

CHAPTER 4.4

GUIDE TO PRESENTATIONS IN THE WORKPLACE

Introduction

Earlier in this book we offered guidelines for simple classroom presentations. Workplace presentations are different from classroom presentations; audience members often have money on the line, and speakers are expected to make recommendations for spending some of that money. Other factors impact workplace presentations as well; audiences do not necessarily know much about your subject before you begin speaking, and they are often pressed for time. These audience members want you to use their time efficiently. This means they want you to inform them quickly about your topic, they want you to sharply define any actions that you want them to take, and they want you to give them enough information to feel that their actions will be appropriate and reasonable. In this chapter we will give some tips on how to design a presentation that is brief and that speaks efficiently to the non-expert members of your audience. We will begin by outlining some strategies for thinking about time and audiences, and we will explain how those strategies govern the organization of an effective technical presentation. We will display and discuss a brief sample presentation, and we will finally explain how to avoid some common and simple problems of slide design.

Planning your presentation

Time

The key to success in a presentation is to recognize that you must be brief. In the (usually short) time allotted, you will be unable to give a comprehensive discussion of your work, so you should not even attempt to do so. Rather, you should plan to display a small number of results and to make a small number of points during your time, and you should look for opportunities to present additional details during a question-and-answer period following the presentation.

Your results, of course, are displayed on slides as diagrams, plots or tables, and you will require a certain amount of time to describe the images on these slides, usually up to one minute per slide. Several factors impact the amount of time you should allow for presenting slides. Plots and tables, for example, can often be described relatively quickly, as they usually display trends or numerical results that can be described concisely. Drawings and diagrams often display more details and raise practical concerns about orientation and about integration with existing systems; these issues can require a speaker to spend more time describing such images than they might spend describing a plot or table. Consequently, speakers whose results are presented as drawings must sharply limit the number of images they will display; a speaker whose results are primarily captured in tables commonly has more flexibility in managing time during a talk.

Goals

In order to plan an effective workplace presentation, you need to set goals for your audience and for your project. That is, in order to assemble details effectively, you need to decide what action you want the audience to take after you finish speaking and how that action should impact the work you and your team are doing. The goal statement need not be elaborate; you should be able to state it in a simple sentence, such as one of the following:

- Your audience should approve your request for funding;
- Your audience should approve your request for additional time on the project;
- Your audience should understand a new job assignment that you've given them, and they should know what steps to take first.

You will set different goals for different stages of your projects, of course. Your goals may also vary according to your seniority; the goal statements above make requests and assignments that are more appropriate for a project leader than for a new member of a project team.

Presentations do not always involve funding requests or job assignments. Many workplace presentations simply disclose information that audience members need to have in order to make plans and to meet deadlines. These presentations are usually called project status reports and administrative policy statements. Project status reports integrate a review of a project's schedule with a summary of current accomplishments, problems and plans. This information allows team members to adjust their work in order to accommodate progress being made on other sections of the project. There are often questions at status report presentations, as team members try to determine how their tasks are impacted by progress on other tasks.

Another type of information disclosure is the administrative policy presentation. Here a speaker explains adjustments that the company is making to the rules—governing matters ranging from network security to the reporting of travel expenses—that employees must follow. Policy presentations need to present clear action descriptions for the audience, as listeners will need to understand what steps they must take in order to successfully comply with new rules.

Your Audience

It is hard to make generalizations about workplace audiences, as your listeners have different amounts of seniority and different kinds of training. However, you can make two simple and powerful assumptions about colleagues and supervisors who are not members of your daily task team:

1) Your listeners are not paid to know everything that you know about your task,
2) Your listeners cannot remember everything you have explained to them in previous presentations.

Your colleagues generally cannot track the details of your project tasks while also making progress on their own tasks. They are too busy to do this successfully. Rather, they will rely on you to give them a clear overview of your task in each presentation and to follow that overview with details describing the interface between their tasks and yours. Consequently, your audience analysis should avoid efforts to define what your colleagues might remember about your work before the presentation starts. Rather, you should define what colleagues should do when your presentation has finished. Generally this means you should organize your presentation to describe your project goals, your current status, your completion schedule and the way your work will interface with the work of your colleagues.

Presentation Organization

A presentation should be viewed as an oral-visual summary of a written report. Consequently, it should be organized according to the same plan that is used in a written report, with some adjustments to accommodate time and audience constraints—it should present less detail than is found in a written report. To respond to audience needs while summarizing a full written report, a presentation should be subdivided into sections as follows:

1. Cover page
2. Outline
3. Introduction
4. Results
5. Discussion
6. Recommendation and Conclusions
7. Questions

This outline and the guidelines below are designed to describe a presentation that is ten minutes long. The example report that is presented in the next section is also designed to represent a ten-minute presentation.

Cover Page

The **Cover Page** of a presentation does the same job as the cover sheet of a report—it identifies the presentation's topic, the author(s), the author's professional affiliation and the date of the presentation. This information requires no comment from the speaker, but because it identifies the speaker, the topic and the occasion, it should be on display when the speaker is introduced,

Outline

An **Outline** is the Table of Contents for a presentation; it shows the topics you will introduce and it shows how these are related. In presentations of 10 minutes and longer, you should display an outline as you introduce your topic.

Introduction

The **Introduction** section of your presentation should be broken into three chunks, and it should fit on two or three display slides. The first chunk should present a Background statement, which should name the topic of the presentation and the problem you are addressing. When you are working for a client, this section should review the client's needs and constraints while listing any other details that may be pertinent to your project work, such as the age of a machine that you must fix, the size of a component that you cannot remove and the like. The second chunk of the Introduction should motivate the work by explaining why the problem is important and by explaining the consequences of inaction. The last part of your Introduction can be called a goal statement or an objective statement; here you should specifically state the actions you are taking to address the client's problem or need, and you should indicate what results you will describe during your presentation.

In our sample presentation, the Introductory information fills five slides, using a Cover Page, a single Outline slide and three slides for an Introduction section.

Results

The second large section of a presentation describes your accomplishments on your project. This section of your talk presents either the **Results** of your work on a problem or, if the work is ongoing, it should present the **Status** of your task, describing work that has been completed to date and comparing this work to your schedule for completion.

Regardless of whether your task is complete or not, you should rely on figures such as drawings, diagrams or plots, to represent your work. If you are delivering a status report on a large task, you should present illustrations to represent any work that has been completed to that point, and you should augment these with displays of your project schedule in order to demonstrate how effectively you are meeting the schedule for your task.

In our sample ten minute presentation, the Results are presented using only two slides.

Discussion

The third large section of a presentation can loosely be called the **Discussion**. Under this heading you should offer explanations or warnings that are significant to the audience's understanding of the results that you have presented. These explanations might include descriptions of alternative solutions that have been rejected, analyses of error or of safety, evaluations of your project's schedule, and the like. The Discussion section of a talk should be brief; you should focus on only two or three significant concerns that your work might raise. Other concerns can be deferred to the question and answer session that follows the formal portion of your presentation.

The size of a Discussion section can be deceptive. The Discussion section of our example report is limited to two tables, shown on two slides, yet these tables represent a significant amount of analytical work.

Recommendation

The next section of a presentation can be called the **Recommendation** section. Recommendations can be presented in a variety of ways, according to the project. When tasks are to be distributed among team members, recommendations are presented as *Action Items*. For the design or purchase of components, recommendations can be presented as *Specifications*.

For many projects, the recommendations can be presented in a statement of a single line. When this happens, it is reasonable to combine the recommendation slide with a Closing or Summary slide, as is the case in our example presentation.

Summary

The last section of a presentation is a **Closing** or **Summary**. This should be presented as a single slide that aggregates the most important points from your Results and Discussion sections. It is common for Summary slides to combine numerical data with action recommendations.

In the example presentation below, the closing Summary is confined to a single slide.

Questions

The last two statements of any presentation should be these: "Thank you. I will be happy to answer questions." These statements establish that your talk has concluded, and they announce the beginning of the unstructured Question session. Circumstances will dictate the length of the question session, although these tend to be brief.

While the question session is unstructured, it presents you with an opportunity to present information that you could not fit into the main section of the talk. To take advantage of this opportunity, you should prepare slides to present any data and analysis points that you would like to explain further, and you should prepare responses to questions that you think your formal presentation might raise.

In responding to questions, you should do two things:

1) Repeat questions after they are asked. This helps other audience members to hear the questions, and it buys you time to organize your response;

2) Locate and display pertinent slides before you begin your answer.

When you do these things consistently during the question session, you will help your audience members by giving effective answers, and you will help yourself by controlling the tempo of the discussion.

CHAPTER 4.5

EXAMPLE OF A DESIGN PRESENTATION FOR A WORKPLACE PROJECT

Introduction

The following is a typical engineering presentation. It describes the results of a stress analysis performed on a component. The presentation is divided into the following sections:

> 1) **Introduction**
>> **Background**
>> **Goal**
> 2) **Results**
> 3) **Discussion**
> 4) **Closing Summary**

In addition to these information sections, the presentation includes a Cover Slide and an Outline slide, both of which are presented before the Introduction slides.

This presentation was limited to 10 minutes with time for questions after the formal presentation had concluded. You will note there is no Recommendations section. This is because this presentation described a completed analysis project; there was no follow-up work planned.

Overview of sample presentation
Section 1: Title and Organization of the Presentation
Two Slides

Fiber Reinforced Polymer Splice for Expansion Joints Colin MacDougall, PhD, PEng January 20, 2010	**Outline** • Introduction to Expansion Joint Splices • Finite Element Model • Stress Results • Discussion of Size and Material • Conclusion
1. Cover slide.	2. Outline.

313

Section 2: Introduction
Three Slides

Expansion Joints

- Allow for expansion of bridge due to temperature
- Subject to severe fatigue loading due to truck loads

1. Background.

Splicing Beams

- Need to connect beams in expansion joints
- Welding beams can lead to fatigue cracks

2. Problem.

New Splice Design

- Company proposing to splice beams using a glass fiber reinforced polymer (GFRP) box
- Avoids welds
- GFRP is fatigue resistant
- Require stress analysis

3. Goals and challenges.

Section 3: Results
Two Slides

Finite Element Model

- 10-node modified quadratic tetrahedron (C3D10M) elements
- Pressure loads to represent truck wheels
- GFRP modeled as an orthotropic material

1. Modeling approach.

Results - Stresses

- Large stresses at GFRP to steel connection
- Potential location for fatigue cracks

2. Results.

Section 4: Discussion
Two Slides

Beam		δ_{max} (mm)
Intact beam		1.515
Spliced beam (using GFRP E= 120 GPa)	L_s = 200 mm	1.982
	L_s = 300 mm	1.741
	L_s = 400 mm	1.484

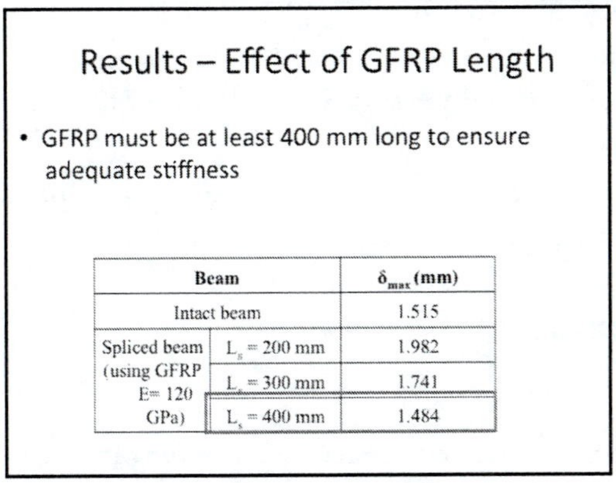

1. Analysis of length.

Type of FRP vs. max. stress developed in the splice			
σ_{max} (MPa)		GFRP (E=120; t=6; L_s=400)	CFRP (E=400; t=2; L_s=400)
Region of splice where max. Stress (σ) is calculated	Midspan	140	168
	End of FRP section at its junction with steel beam	167	327

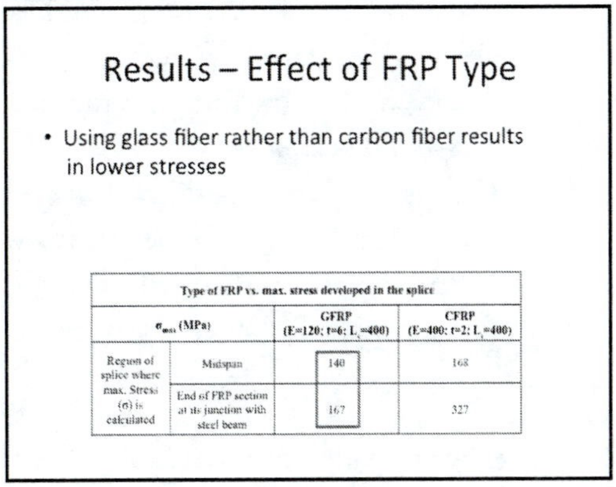

2. Analysis of type.

Section 5: Closing Summary
One Slide

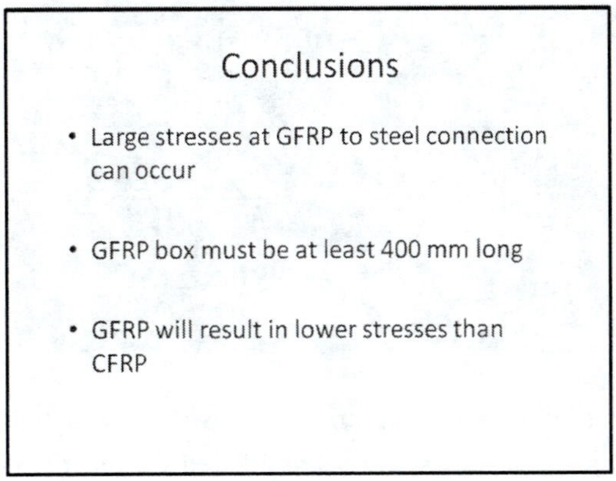

Planning your slides

Before preparing your slides in detail, it is important to prepare an outline of the presentation:

1. Start by defining how much time you have to speak. This will allow you to determine approximately how many slides to use. *A good rule of thumb is "one slide per minute."*
2. Once the total number of slides is defined, determine the number of slides available for each of the major sections of the presentation. Usually three slides will be taken up for the Title, Outline, and Closing Summary. The remaining slides need to be proportioned between the Introduction, Results, Discussion, and Recommendations. *A common mistake is to create far too many slides and spend far too much time on the Introduction section of a presentation.*

3. Once the number of slides available for each section has been defined, you can identify the critical ideas you want to convey to your audience on each slide. *Start by identifying the two or three critical results.* Decide how you want to display those results (usually a figure, graph, or table). Do the same for the Discussion and Recommendations. Finally, outline the slides for the Introduction and Closing Summary. *Don't worry at this point about how the slides look.*

4. Once you have an outline of the presentation, run through the presentation from start to finish to ensure the ideas flow logically.

5. When you are satisfied with the outline of the presentation, you can then work on formatting each slide so that it clearly presents its information to the audience.

Title

The Title slide provides the title of the presentation, the name and credentials of the presenter, the affiliation of the presenter, and the date of the presentation.

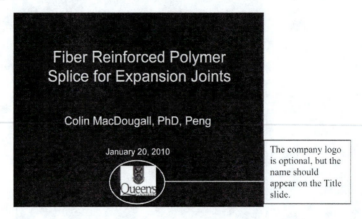

Often there is more than one contributor to the work that is being described by the presentation. In this case, all contributors are usually listed on the title slide, with the presenter listed first:

Outline

The Outline provides the main headings for the presentation. There should rarely be more than 5 or 6 main headings for the presentation. When presenting the Outline slide, do not go into details. This wastes time, and the details will come later in the presentation.

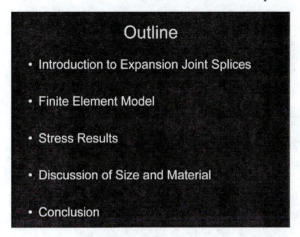

While you should not give details in your outline, you should give specific information about your presentation. You can do this efficiently by showing topics for each of the general section headings. Here, for example, the vague heading "Results" is changed to the specific "Stress Results," and the vague "Discussion" heading is modified to add the specific topics of Size and Material.

Introduction

The purpose of the Introduction is to help the audience appreciate why this problem is important (Background), and to define the particular aspect of the problem you are investigating (Goal). The Introduction is also an opportunity to define terms that will be used later in the presentation.

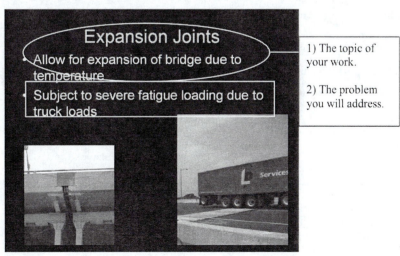

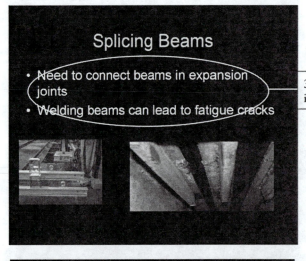

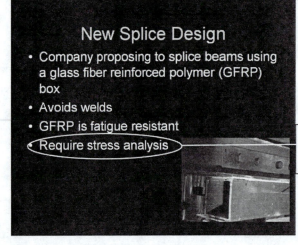

An Introduction must be efficient in order to save time. To introduce your work efficiently, you should concentrate on four statements that the audience must see:

1) The topic,
2) The problem,
3) The importance of the problem,
4) The task you performed to address the problem.

Information that does not serve these statements is likely to distract your audience.

Results

You should begin a Results presentation by describing the method used to obtain the results. You will likely not have enough time to provide all the details on the methods during the presentation. Often a report is written in conjunction with the presentation. The report should contain detailed information on the Methods. A good tip to save time is to state that:

"Detailed information on the methodology has been provided in the final report." Of course, be prepared to answer questions if they arise (it is often worthwhile to have additional slides prepared to help answer these questions).

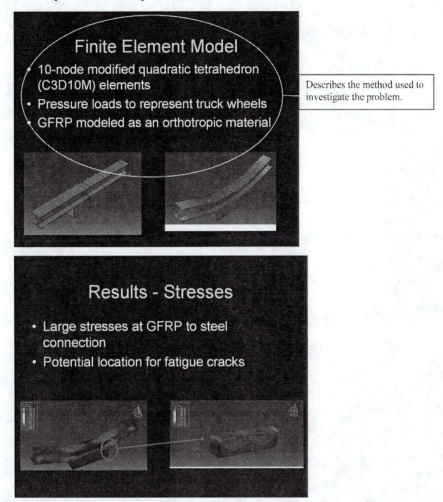

You will likely not have enough time to display all your results during your presentation. Therefore, narrow your focus to the two or three most important results.

Discussion

For this presentation, the Discussion involves two tables. Tables often contain too much information for the audience to absorb. In the slides below, red boxes were used to draw the audience's attention to the most important points in each table.

Effect of GFRP Length

- GFRP must be at least 400 mm long to ensure adequate stiffness

Beam		δ_{max} (mm)
Intact beam		1.515
Spliced beam (using GFRP E= 120 GPa)	$L_s = 200$ mm	1.982
	$L_s = 300$ mm	1.741
	$L_s = 400$ mm	1.484

Effect of FRP Type

- Using glass fiber rather than carbon fiber results in lower stresses

Type of FRP vs. max. stress developed in the splice			
σ_{max} (MPa)		GFRP (E=120; t=6; L_s=400)	CFRP (E=400; t=2; L_s=400)
Region of splice where max. Stress (σ) is calculated	Midspan	140	168
	End of FRP section at its junction with steel beam	167	327

Closing summary

In the Closing Summary, display again the most important two or three points from the presentation. *Do not introduce any new information not discussed during the presentation.*

Chapter 4.6

Guide to Performance During Presentations

Introduction

Presentations differ from written documents in that you are on display for your audience at all times, and your actions can have an impact on the audience's response to your words. There are several things that you can do that will assure that your audience understands what you are saying:

1) Rehearse the presentation.

You should speak through your presentation fully before you appear before your audience. It is best to rehearse in the room where your talk is scheduled, as this will allow you to get accustomed to the projection equipment that you will use. If you have no access to that room, however, you can simulate the presentation by advancing slides on your own computer. This will help you to determine specifically how you will explain your figures and tables, and it will help you to orchestrate the small steps involved with advancing slides. If you are presenting with a partner, you should rehearse with that person; it is important for each member of the team to hear what the other will say, and it is important to know how much time your partner needs in order to complete his or her section.

2) Make Points as you speak.

Your slides present information that the audience needs to see and remember. Audiences understand presentations best when you can state the takeaway information—the point—of each slide in a single sentence. You should develop that point sentence as you prepare each slide. An easy way to do this is to complete this sentence in your notes for each slide: "From this slide, my audience needs to remember [*blank*]." The statement in the bracket should fill one simple sentence. If the point is longer than one sentence then your slide may be cluttered with unrelated ideas.

3) Use a pointer.

When you describe drawings or tables, you should direct the audience's attention to particular parts of those displays while you describe them. For this you need a pointer. When you point at the screen, keep your pointer still; when you move your pointer in large circles, audience members tend to lose focus on the components you are describing.

4) Face the audience as you speak.

You will naturally look at your displays as you move your pointer, and you will look at your projector as you change slides. However, you need to make sure that you turn your head towards the audience during most of the time when you discuss each slide.

5) Prepare handouts.

When you give a presentation, you should provide a handout to your audience. They might use this to take notes or they might review it later to remind themselves of your points. Speakers often distribute copies of their slides as handouts. This is acceptable, but it may not provide your audiences with adequate information, particularly if you have followed our advice to sharply limit the words on your slides while relying on figures to present your information. You should also prepare a one-page handout that captures the points and explanations that you have omitted from your slides, while aggregating your two or three most significant results displays.

The logistics of slide presentations

Even at small businesses, people are expected to use computer projection systems to give presentations, so you need to understand how these systems work before you prepare your talk. While we cannot predict the particular set of plugs and buttons that you will encounter each day, we can outline some of the constraints that business projection systems impose on users so that you develop a default presentation that is robust to the most common problems. Here, we speak not about the computer systems so much as we speak about the projectors and screens that actually display the information you load into your computers. We begin by speaking generally about the problems of screen space and light. After this we will discuss the problems of slide design and the way you can use slide space to clarify information and to acknowledge your affiliations.

Screen Space

Most business projectors use a 4 : 3 aspect ratio. This means that they cast light in a rectangle that is wider than it is tall. Narrowly, this means that you should choose *landscape* orientation when you review the settings for presentation slides or pages. Additionally, because the letter-size page to which many programs default does not have 4 : 3 proportions, projectors may resize images in a way that degrades your display. PowerPoint solves this problem automatically, as its default slide, measuring 10 inches x 7.5 inches, uses this 4 : 3 ratio. If you design your presentations using other programs, you will need to verify that your page settings will fit a similar 4 : 3 frame without distortion.

Light

Presentations depend on light, and light can be unreliable. The source of light for your presentation is a projector's lamp; your eyes then receive that light by reflection from a screen. The lamp and the screen can both be sources of distortion for your displays; projector lamps fade as they age, and screens lose reflectivity as they accumulate dust. When this happens, colors can fade and distinctions can be obscured. Light infiltration from windows or open doors can create additional distortions that are hard to control.

While you cannot control the projection equipment that you will use, you can plan for worst-case situations by developing slides that display well in poor light conditions. Usually this means reducing the amount of color you use in your slides; to the extent possible, you should exchange dark or patterned backgrounds for light, solid colors and exchange color illustrations for black-and-white line drawings. You need not eliminate color from your slides; however, you should select colors in such a way that your information will display well even if the colors fade towards gray. In practical terms, this means that line drawings tend to display more reliably than solid model drawings, and that photographs usually display poorly unless they are heavily annotated or edited. If you wish to edit a photograph to enhance contrast for display, you must disclose your editing steps to the audience, usually by displaying the unedited image before presenting an enhanced version.

Responsibility

When you give a presentation, you should always display a cover sheet showing your name, the names of your team members, the title of the work and the date of the presentation. The presentation date should be visible on your slides as well. These displays allow your immediate audience to spell your name correctly, of course; more importantly, however, these define who is responsible for the accuracy of the information that is being presented. As a professional, it is important for you to do this because your slides are likely to be transmitted to people who did not attend the presentation, and they may be reviewed years after you have left the project. The people named on the slides should be ready both to receive credit when work is well done and to answer questions should problems arise later.

You should also list the names of your company and/or your sponsor on the cover slide, either with a text entry or with a corporate logo. Such listings provide a professional acknowledgement for those who may have funded your work, and they disclose the parties who may claim ownership of any intellectual property you may have created.

Slide geography.

Presentation programs such as PowerPoint use a default slide size of 10 x 7.5 inches. When your projector and screen are placed correctly, such default slides will exactly fill your screen. However, slides are typically subdivided into a central display block and a border, as suggested in Figure 1.

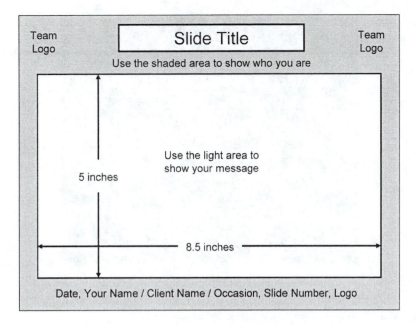

Figure 1. Space arrangement of a typical 10 x 7.5 inch slide

In Figure 1, the light-colored areas at the center represent space that you can use to display your information. The shaded areas at the perimeter of the slide represent space that is commonly reserved for identity information—authorship, date, team and/or corporate emblems and the like. Some companies will require you to use a pre-defined company slide that has a wide border for corporate emblems; other companies will give you freedom to enlarge the display space while minimizing the corporate branding displays.

Displaying Information

In a presentation, you provide two kinds of information—words and images. Words can be displayed in the form of text slides, labels on drawings and column heads in tables, while images include drawings, diagrams, plots and photographs. Usually your slides will present a mixture of words and images, and your challenge is to design that mixture in a way that is clear and that displays effectively. Handy rules of thumb are these:

1) Use type fonts between 20 and 40 point.

In most rooms, audience members will be unable to see words that are smaller than 20 point in size. This size restriction limits the amount of text you can display on a screen; you should avoid complete sentences in favor of key-word or phrase reminders. If your points require extended text discussion, you should place that discussion in a handout that audience members can review later. During the talk, it is better for them to listen to your explanation than to read your slides.

Figure 2 presents an example slide that uses 40-point Arial type throughout.

OBJECTIVES

• Model response of trackwork to passage of a vehicle

• Predict stresses and deflections of typical cross-tie under service loads

Figure 2. Text slide with 40 point type

2) Design images to use all of the available display area.

In presentations, plots, tables and drawings should be large. You should prepare your visual information so that it will fill the display space on your screen as in Figure 3. Images are used to convey information that cannot be conveyed efficiently in words; they capture something crucial about your work, and you need to make sure that your images are large enough to display that crucial information sharply.

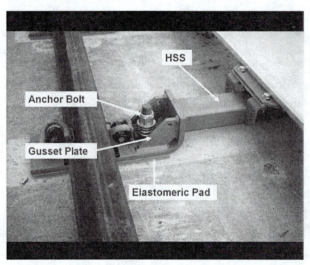

Figure 3. Images should fill the screen

In Figure 3, even though the components are well labeled, they occupy only a small portion of the image, and their color differences are small. This image needs to be large in order for the bolt, plate and pad to be easily visible.

When you display an image in an oral presentation, as in a written report, it is important for you to speak to the figures in detail. To do this, you should

1) Introduce the figure by specifically naming the device or component that is on the screen;
2) State why the figure is being shown by defining the point it makes or the result it presents;
3) Call out the labeled components of the display;
4) Discuss the display, explaining how it works or how you will use it in your work.

3) Minimize words on slides with figures.

Speakers commonly mix words with figures. This is a matter to approach with caution; in some circumstances words can enhance a figure display, while in other circumstances they create more problems than they solve. In general, words can usefully be applied to an image in the form of slide titles or labels for important features of the image. Such labels should be descriptive, short, and they should be placed near the features they describe; slide titles should naturally fill the border space at the top of the slide. Figure 4 is an example:

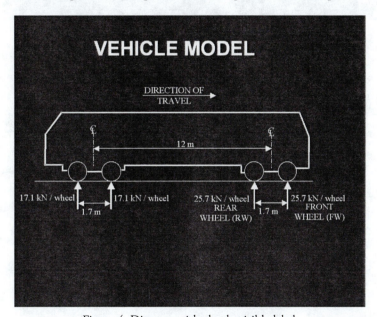

Figure 4. Diagram with clearly visible labels

Text creates problems for you when it is used to explain or comment on an image that is already labeled. Discursive statements are often lengthy, and they occupy valuable slide space; they may well obscure the components that are being evaluated.

Figure 5 is an example of a poor mixture of text and image, as the words and the image occupy roughly equal amounts of slide space. This is unacceptable; the text block uses a large amount of space to present trivial information, while the plot is so compressed that distinctions between the two data sets are rendered invisible.

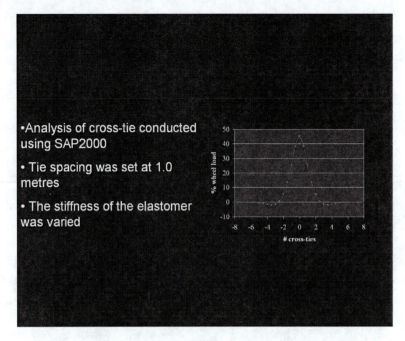

Figure 5. Side-placed text reduces space for displaying images

If commentary must be displayed on an image slide, it must be short, and it must be placed on an unused area of the slide, such as a border area or, in the case of an irregularly shaped image, on an area that remains vacant after the image has been sized. In Figure 6 below, the plot has been resized to fill the slide, and the side-placed text has been removed. Labels are used to convey critical information that is often presented in text bullets. Because images are commonly wider than they are tall, some text information can usually be placed above or below your display in the form of a slide title or a legend.

In this configuration distinctions between the data sets become visible, and this slide can stand on its own if circulated by members of your audience.

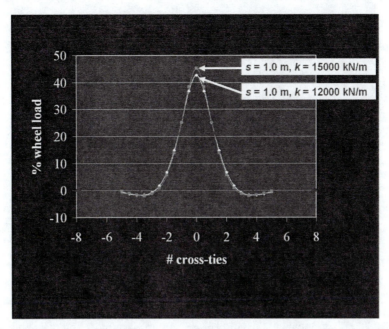

Figure 6. With text removed, the plot is resized and critical information is inserted in label form.

Slide designs and backgrounds

Presentation programs such as PowerPoint make it easy for users to design slides with richly colored backgrounds. These programs come with libraries of slide design schemes and with tools that allow users to develop and share alternative backgrounds. Some of these backgrounds support good presentations and others do not. Our purpose here is not to impose a particular design and color scheme on your slides but to outline the general issues you should consider as you begin to develop your own presentations or to develop a template that your company might use widely. To do this we will present several commonly used slide designs, and we will discuss the strengths of each and the problems that they present.

Color

Slides communicate information to audience members' eyes. Because light conditions can be poor, it is best to prepare slides with high contrast between the background and the information that you want your audience to see and remember. The simplest way to obtain high contrast slides is to place dark text and lines on a white background. Black-on-white slides are easy to prepare, and they are of sufficiently high contrast to display effectively even under poor lighting conditions. Such slides also reproduce well as handouts; this is an important matter, as presentation slides are often printed or copied using black-and-white office machines. The example black-and-white slide in Figure 7 was first introduced in an earlier chapter:

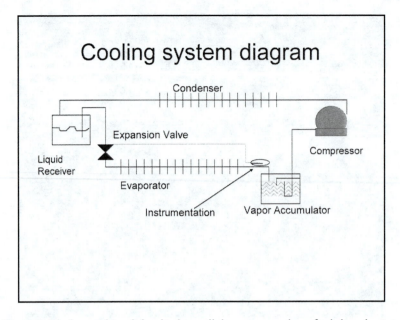

Figure 7. Black-on-white slides display well, but many authors find these boring.

While black-on-white slides are excellent for reproduction, many authors find them to be dull, and some audiences find white backgrounds to be uncomfortably bright. To address these problems, many professionals invert their color schemes, using blue backgrounds for light-on-dark displays. When slides mainly present text, such backgrounds work nicely; however, diagrams and drawings tend not to display well on dark backgrounds, as indicated in Figure 8:

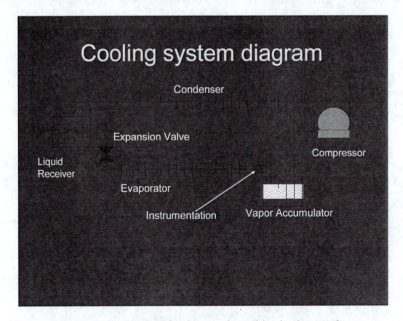

Figure 8. Dark colors display poorly on blue backgrounds.

Here we have reproduced the cooling system diagram from the previous slide, and the slide background here clearly obscures the diagram. One can, of course, re-color the lines in this diagram in order for them to display well against the dark background, as in Figure 9:

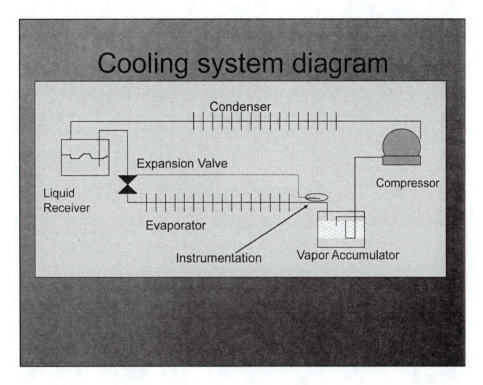

Figure 9. A diagram is revised to accommodate a slide's background color.

Here the author has altered the technical display in order to accommodate the dark background, and this is a mistake. In addition to taking on unnecessary work, the author has trivialized the substance of the presentation by making it serve the presentation's color scheme. When you have the opportunity to choose, you should adjust the slide background rather than your professional output. This will save time for you, and it will make it easier for you to reproduce and adjust your slides later.

You can enhance a presentation without distorting your information. Presentation programs will allow you to develop your own background colors quickly and easily. Generally speaking, the most robust backgrounds will balance the color's hue, intensity and brightness. The example slide in Figure 10 shows a background that was developed using this principle:

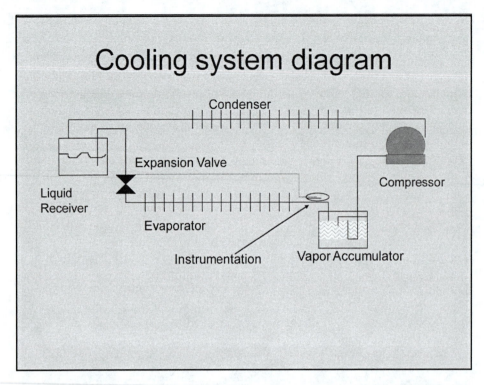

Figure 10. Robust slide backgrounds should be light in hue intensity and high in brightness.

The slide is still blue in hue, but the color intensity is low, and the color is relatively bright. By adjusting these factors, you will be able to explore a wide range of background colors, many of which will give you satisfactory results as presentation backgrounds.

Patterns

Many authors avoid solid-color backgrounds in favor of patterns, explaining that solid colors leave distracting empty space on slides. Presentation programs offer such authors patterned and decorative backgrounds to aid in the visual design of their slides. However, as was the case with color backgrounds, patterned slide backgrounds can present as many problems as they solve, and they should be used with some care. Figure 11 is an example of a blue slide to which a gradient color effect has been added such that the slide darkens in a diagonal from the top left to the bottom right.

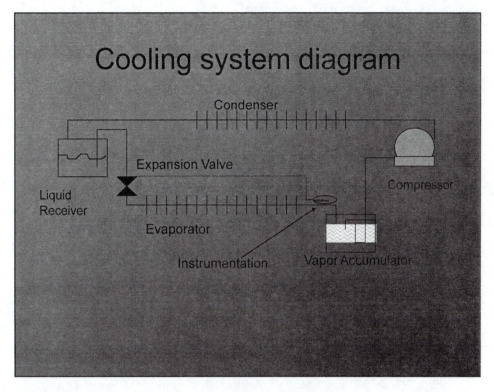

Figure 11. Gradient color schemes can generate an array of contrast problems.

While gradient slides such as this one are commonly used, they can compound the problem of color contrast that we discussed earlier. With a gradient background, authors must work harder, because they must review the contrast that their displays form with several distinct color areas on their slides.

Background Images

As an alternative to gradient backgrounds, authors may select from a variety of pre-bundled background images. One of these is shown in Figure 12 as part of a slide that again displays our cooling system diagram; it should be clear that several things are amiss. The background image is unrelated to the diagram being presented, and this creates a distraction; if the audience is paying attention, they are likely to expend some energy considering this disconnect. More generally, there is some risk that the background illustration could be more interesting than the diagram presented on the slide, a distraction that you as a presenter do not need to have.

Technical professionals can always take steps to mitigate the problems raised by their backgrounds. In Figure 13, the color problems of the gradient background are addressed by creating a secondary background—a drawing box—for the image. While this trick solves the display problem raised by the decorative background, it does so by resorting to the kind of solid, light display background that was described earlier.

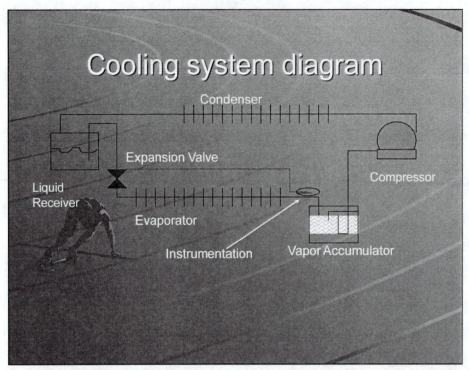

Figure 12. Decorative schemes can be more interesting than your information.

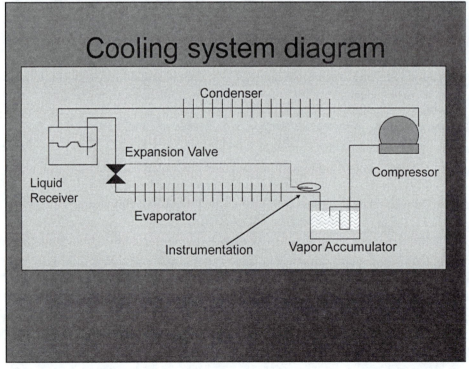

Figure 13. A light display box can solve the problems created by dark backgrounds.

Space

Slide design schemes present authors with numerous problems regarding reasonable use of color and decoration. However, design schemes can also impact the space that is available to display your information. Figure 14 shows a widely-used slide design scheme that places a decorative, non-intrusive splash of color in the upper left side, along with a long horizontal bar near the bottom of the color design. That horizontal bar defines the bottom edge of the slide title box, setting aside the top two inches of the slide as a title box. The slide design is pleasant, but the decoration fills up a great deal of space that could be used to display your figures.

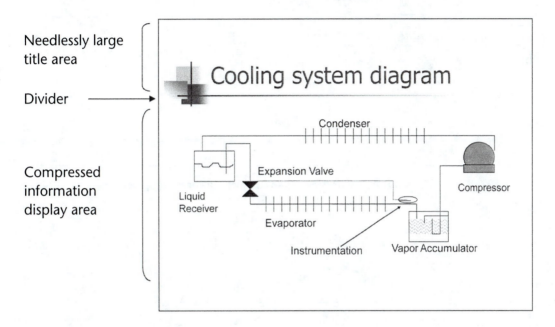

Figure 14. Decorative design schemes can reduce the space available for information.

Some slide designs impact horizontal space as well as vertical space, as in Figure 15. As did the design in Figure 14, this slide decoration devotes two inches to the slide title, and it further imposes a needlessly wide border on the left side of the slide. Our cooling system diagram must be resized in order to fit it onto a slide using this design.

You can, of course, move these horizontal bars and borders by editing the master slide, and you should do so if you wish to display technical information on slides using these patterns. Alternatively, you may select slide designs that restrict decorations to the border area that was described earlier—slide space that you usually keep empty to make room for corporate emblems, and to accommodate misalignment of screens and projectors. In Figure 16, a decorative frame has been imposed just at the edges of the display area. This offers a pleasing band of color while not disrupting the display of information.

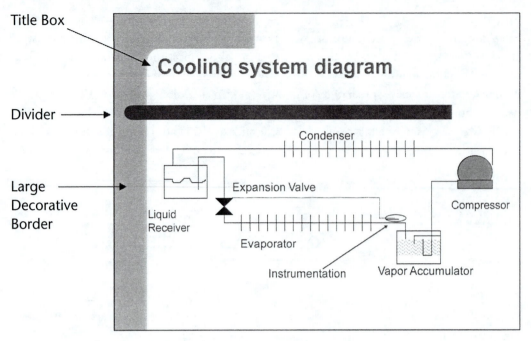

Figure 15. Heavy framing design schemes can constrain space.

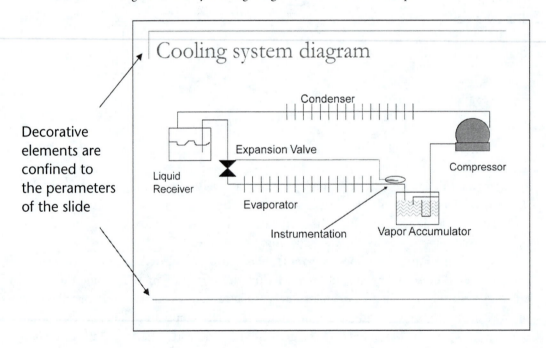

Figure 16. Light framing design schemes are usually acceptable.

CHAPTER 4.7

GUIDE TO MEMORANDA

Introduction

The memorandum is the most common type of business communication. It is the medium used by people within a company or team as they discuss business—particularly to announce policies, assign tasks or solicit input.

Memoranda differ from letters in that they address different audiences. Letters are vehicles for external communication—with clients, for example, or regulators. For these external audiences, letters are used to describe actions that the reader should take or actions that the company may have taken in response to previous requests. Memoranda differ from reports in that reports tend to be oriented to the impersonal presentation and analysis of facts. Reports may present facts that justify action, but the actions themselves are defined and explained in the letters or memoranda to which the reports are attached.

Memos are commonly used as transmittal notes for larger reports, or they may be used to direct readers to other information sources. Memoranda may also be stand-alone documents, as when they are used to announce changes to company procedures or team task assignments.

Memoranda can be partitioned into four information sections. In order of appearance, these are the Header, the Introduction, the Details presentation and the Action Item. These are discussed next.

1. Header

The header is the most distinctive format section of a memorandum, as it displays the well-known block of boldfaced words **To**, **From**, **Subject** and **Date**. These should be filled out fully and specifically, as they display the lines of responsibility for the information presented in the memo, and they display the date after which the recipient can be assumed to have known his or her responsibilities with respect to the subject matter. The subject statement should be fully descriptive; it should name the work-related topic to which the memo refers, and it should name the actions that the memo requires or the changes that it will present.

2. Introduction

The text of a memorandum begins with an Introduction statement that should *name* the topic of the communication, *motivate* that topic and state the *action* the reader will be asked to take. This is a great deal of information to present, but it must be presented very briefly. This means that your introduction should mainly name the primary concepts of your communication, leaving the details for a later section of the memo. In this way, you could say that the Introduction to your memorandum is functionally similar to an Abstract.

3. Details

The middle section of a memorandum presents details in support of—or in explanation of—the action that readers are to take. This section tends to be the longest section of the document, although its length will vary depending on the complexity of the topic and the significance of the actions that readers are asked to take. Those actions should be described briefly but specifically; readers must be explicitly notified of any steps they are expected to take, and they must have enough information to take those actions successfully. You should define actions sharply while keeping the number of those actions small, as in these action statements:

> "Please fill out the attached survey and return it to me by the end of business tomorrow."

> "Please contact Human Resources and review your W-4 as soon as possible."

In memoranda, you should avoid asking readers to perform intricate tasks. If a task involves more than one or two small actions—as when they are updating or installing a computer program—you should provide assistance, instructing readers, for example, to cooperate with a staff member who you might send to their offices to perform these tasks for them.

4. Action Items

At the close of a memorandum, requirements should be defined for all the parties to the communication. These include the actions that you are asking readers to take and the actions that you are taking in order to assure that they are successful. Statements of author and reader actions should be paired. If your readers are expected to obtain further information elsewhere—from your company's website or from a vendor, for example—you should explicitly state how they should obtain that information. Finally, at the close of any action statement, you should offer to respond to your readers' questions or to assist them in, for example, scheduling their responses, determining if they are impacted, or the like. In a workplace—even a medium-sized business—this offer should specify who in your organization will respond to individual queries and how that person is to be contacted.

You should specify a human contact for two reasons. First, the person issuing the memo may not be the person who is best able to respond to questions. Second, and related to the first, if the contact person is not specified, there is likely to be a certain amount of confusion and time lost as people with questions or concerns learn where to direct their queries.

Next we present an example memorandum that illustrates these four information sections in a way that is both concise and concrete. Each line of this document performs a function for the author and the reader, and this brief discussion will call attention to these functions.

The Heading section specifies first who is to read the memo—the staff of *S&J Mechanical*—and who is responsible for issuing the memo—the president of the company. Its date indicates when the audience is expected to have obtained the information in the memo, and the subject line describes the topic sharply. This subject line is of particular importance here, because it concretely names a topic—the company's medical benefits—as well as a significant action—*changes* in these benefits.

The Introduction paragraph presents background information in the first two sentences while naming medical benefits in particular as a topic for further discussion. The last sentence fully names the issue—changes to the company's health benefits—and an action that readers should take—to review these changes in time to take advantage of open enrollment.

The Detail section of the memo makes one significant point about each of the insurance plans mentioned in the opening section. Each of these points is accompanied by a brief evaluation, suggesting possible actions based on the change that is described.

The closing section explicitly indicates what actions the company has taken to provide information to users, and it indicates what actions users can take to obtain that information. This is done by inserting the electronic address of the company's benefits page, where more detailed brochures are available for employee scrutiny. It also lists the name of a contact person who will function as the company's expert on this matter.

CHAPTER 4.8

EXAMPLE OF A MEMORANDUM FOR WORKPLACE DISTRIBUTION

Heading:
Responsibilities
are defined:

* Reader
* Author
* Date
* Topic

Introduction:
Topic
Motivation
Action

Details:
Issue 1
Action 1

Issue 2
Action 2

Issue 3
Action 3

Actions:
Author / company
actions

Reader actions

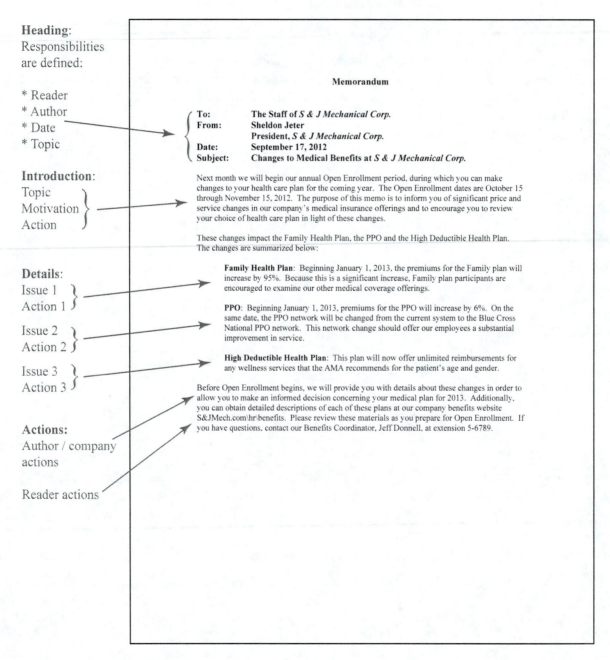

Memorandum

To: **The Staff of *S & J Mechanical Corp.***
From: **Sheldon Jeter**
President, *S & J Mechanical Corp.*
Date: **September 17, 2012**
Subject: **Changes to Medical Benefits at *S & J Mechanical Corp.***

Next month we will begin our annual Open Enrollment period, during which you can make changes to your health care plan for the coming year. The Open Enrollment dates are October 15 through November 15, 2012. The purpose of this memo is to inform you of significant price and service changes in our company's medical insurance offerings and to encourage you to review your choice of health care plan in light of these changes.

These changes impact the Family Health Plan, the PPO and the High Deductible Health Plan. The changes are summarized below:

Family Health Plan: Beginning January 1, 2013, the premiums for the Family plan will increase by 95%. Because this is a significant increase, Family plan participants are encouraged to examine our other medical coverage offerings.

PPO: Beginning January 1, 2013, premiums for the PPO will increase by 6%. On the same date, the PPO network will be changed from the current system to the Blue Cross National PPO network. This network change should offer our employees a substantial improvement in service.

High Deductible Health Plan: This plan will now offer unlimited reimbursements for any wellness services that the AMA recommends for the patient's age and gender.

Before Open Enrollment begins, we will provide you with details about these changes in order to allow you to make an informed decision concerning your medical plan for 2013. Additionally, you can obtain detailed descriptions of each of these plans at our company benefits website S&JMech.com\hr\benefits. Please review these materials as you prepare for Open Enrollment. If you have questions, contact our Benefits Coordinator, Jeff Donnell, at extension 5-6789.

CHAPTER 4.9

GUIDE TO E-MAIL IN THE WORKPLACE

Introduction

E-mail communications are formally similar to memoranda, with some modifications driven by the user's interface with the mail program. In the most significant such modification, the memorandum heading block is replaced in e-mail messages with electronic address and subject fields; dates are automatically assigned at the time the messages are sent. When an e-mail message is viewed onscreen, these fields are commonly displayed at the head of the message, arranged as heading blocks that have a different appearance but a similar function to the heading section found in a memorandum. This memorandum layout is typically retained when e-mail messages are printed.

Like memoranda, workplace e-mail messages should be oriented towards action. Specifically, in business e-mails you should issue instructions, confirm instructions that you have received, or you should respond to questions that others have raised. You should expect your e-mail messages to become part of your professional documentation, which means that they may be preserved with your administrative files or your project records. As a result, your e-mails need to characterize, for example, the issue—question or problem—you are responding to as well as your answer. Additionally, e-mails need to record the date when the issue arose, and they should indicate who brought you into the discussion. This context information should be introduced—or preserved—to document the background of each of your communications.

E-mail is widely used for both private and professional purposes, and you should work carefully to keep personal e-mail separate from your professional e-mail. The best way to do this is to maintain separate e-mail accounts, using one for your business-related correspondence and the other for private communications. This helps you to avoid the serious problem of using your company's system to store or process sensitive or personal information, and it helps you to avoid embarrassment that may arise if your company's address list should become mingled with your personal address list.

Most people receive an enormous volume of e-mail. Much of this is junk that eludes the company's filters, some of it is workplace mail that has been distributed more widely than necessary, and some of it is unnecessary chatter among colleagues. To prioritize their mail, most users have developed two characteristics that greatly complicate workplace use of e-mail. The first of these is *impatience* and the second is *informality*.

1) Impatience

Users have adapted to the high volume of e-mail by learning to skim their inbox headers seeking familiar **Senders** and pertinent **Subjects**. Many people open and read messages only after a message meets their Sender/Subject criteria. They then read quickly, searching for motivation, context and action information; readers may not finish reading notes that do not present that information in the first few lines.

As a result, your messages are screened twice—once from a mailbox-level scan of the Sender and Subject, and then again from a scan of the opening paragraph. For your mail to register on your reader, you need to make sure that these sections of your message are specific, direct and sharply stated. As with memoranda, the subject line must specifically define the topic/action of the message. Then in the first lines of your message, you should define why you are writing and what you want your reader to do. You should defer details about these matters to the body of your message.

2) Informality

E-mail is a common tool for informal communication. Such communication is generally conducted for fun rather than for transacting business, so the participants do not work hard to be specific, or to define actions and responsibilities. Rather, when people write informal e-mails quickly, they use nicknames and they avoid specificity in naming topics and actions. Workplace standards, however, are different from informal standards; workplace notes need to be specific, concise and accurate. When authors accidentally compose business-related e-mails using informal standards, the result is usually unprofessional and confusing. Most informal e-mails will be ignored because they fail to present enough information for the reader.

Many of the problems related to informality can be addressed by making adjustments to your e-mail settings in order to assure that author and recipient names are fully displayed with each message. The display of recipient names can be managed by making adjustments to your electronic address book, which you can edit to display first and last names along with electronic addresses. Display of your name should be addressed in two places. First, you should assure that your first and last names appear in the "sender" column of your recipient's e-mail display without nicknames. Usually you can do this by adjusting the identity setting in your e-mailer's options or preferences menu. Additionally, you should take advantage of your mailer's ability to insert a signature at the bottom of each e-mail message. Your signature should display your full name, your job title and your mailing address at your company, followed by your work telephone number. You should restrict your workplace signature file to official contact information; avoid aphorisms, quotations and artwork.

While you can automate your e-mail program to manage the identity displays, your **Subject** field requires thoughtful input from you each time you compose a message. The subject line for an e-mail message performs the same job as the subject line for a memorandum; however, because readers may prioritize—or even dispose of—e-mails on the basis of the subject statement, your subject must sharply characterize both the topic of your message and the action you want the reader to take. When possible you should design subject statements that define first the project and second the action that is to be taken. The following example subject statements demonstrate how descriptive topic and action phrases can be combined to make useful subject statements:

Subject: City Hall Renovation – request for reimbursement

Subject: Atlanta Tower Crane – maintenance required

Format in e-mail

After the specifics of subject and sender have been addressed, the e-mail format resolves into a compressed version of the Memorandum, with an orienting Introduction, a detail-rich Discussion, and a brief Closing that focuses on actions the reader should take.

1) Introduction

The **Introduction** section should include a salutation line like those used in print letters, followed by an Introduction statement similar to the Introduction in a memorandum. This Introduction should compress the topic announcement and objective into a single statement of the *purpose* or *goal* of the message. If your message is prepared in response to a request from a third party—your supervisor, for example, or a client—you should introduce this as background information. The following examples suggest how an e-mail might combine a purpose statement with a request response:

Dear Bob,
Sheldon Jeter has asked me to respond to your query of June 23 concerning…..

Dear Stephen,
I am writing in response to a request you posed to Sheldon Jeter regarding….

2) Discussion

A Discussion section should answer questions and explain the answers, or it should itemize the actions the reader is expected to take, offering brief explanations of those actions, as necessary. In workplace communications, all explanations must be specific. People—both colleagues in your company and contacts at other companies—should be identified by their full names; contact information such as telephone numbers, e-mail addresses and regular addresses should be included for people who are not employees of your company. Products and purchase orders should be specified by name and identification number, and the dates and substance of previous contacts should be specified when they are cited in the message.

Readers may be asked to take a variety of actions when they read e-mail messages. They may be asked to read and respond to an attached document or electronic link, they may be asked to contact a third party, or they may be assigned a project task. Regardless of what particular action you want your reader to take, you should specify that action, you should define a date by which the reader should complete that action, and you should indicate where the reader should direct the results. The following action statements are suggestive:

> I've attached 1) a manuscript for you to review and 2) a reviewer checklist. Please respond to the questions on pages 2 and 3 of the reviewer checklist and deliver the responses to me in e-mail by Friday the 23rd.

> Please contact Bill Jones, our service representative, at 123-456-7890. Give him a purchase order number for the mass spectrometer package that we ordered and ask him to ship the device. Get a tracking number for the shipment, along with a delivery date. Get back to me before the end of the day with the shipping date and tracking number.

3) Closing

At the close of an e-mail, most authors politely offer to provide assistance to the reader. These assistance statements are commonly pleasant, but they are not always useful, as they provide phone numbers for the reader, but they do not clarify the type of assistance that the reader can obtain:

> Please let me know if I can be of further assistance. If you have questions, you can contact me at the above address, or you can call me at (123) 456-7890.

It is good to offer assistance, and such non-specific offers may be adequate for tasks that the reader is to perform for his or her own benefit. But when the reader is expected to deliver a result to a colleague, more details are usually required in the closing section. In an action-oriented e-mail you should define a specific time for completion of a task, you should define what is to be delivered when the task is completed, and you should define who receives the result or notification at the time of completion. An example closing might be this:

> I will review this using the Track Changes tool, and I'll deliver the edited file to you on Wednesday morning. Please contact me if you haven't received the file by Wednesday at noon.

Penalties for failure can be mentioned in closing sections as well:

> The deadline for submitting reimbursement requests for transactions in the 2008 fiscal year is 5:00 p.m. on June 30. Any requests for FY 2008 reimbursements that are filed after that date will not be processed.

Next we present an example e-mail exchange concerning a project task assignment. This e-mail was used to define the responsibilities of two employees and to set deadlines for completing a project and a project report.

CHAPTER 4.10

EXAMPLE OF E-MAIL EXCHANGE FOR A WORKPLACE PROJECT

Heading:
* Responsibile parties
* Date
* Topic

Introduction:
* Goal and context
* Task assignments

Discussion:
* Details of task
* Responsibilities
* Contact information

Closing:
* Completion schedule
* Offer of assistance

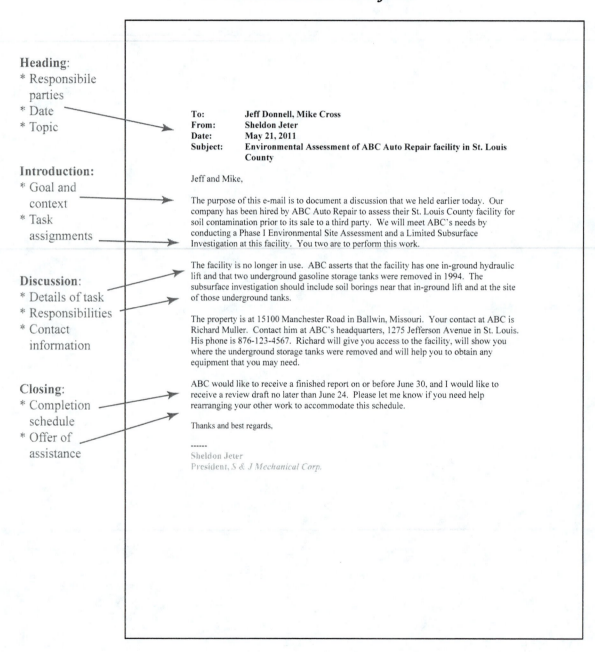

To:	Jeff Donnell, Mike Cross
From:	Sheldon Jeter
Date:	May 21, 2011
Subject:	Environmental Assessment of ABC Auto Repair facility in St. Louis County

Jeff and Mike,

The purpose of this e-mail is to document a discussion that we held earlier today. Our company has been hired by ABC Auto Repair to assess their St. Louis County facility for soil contamination prior to its sale to a third party. We will meet ABC's needs by conducting a Phase I Environmental Site Assessment and a Limited Subsurface Investigation at this facility. You two are to perform this work.

The facility is no longer in use. ABC asserts that the facility has one in-ground hydraulic lift and that two underground gasoline storage tanks were removed in 1994. The subsurface investigation should include soil borings near that in-ground lift and at the site of those underground tanks.

The property is at 15100 Manchester Road in Ballwin, Missouri. Your contact at ABC is Richard Muller. Contact him at ABC's headquarters, 1275 Jefferson Avenue in St. Louis. His phone is 876-123-4567. Richard will give you access to the facility, will show you where the underground storage tanks were removed and will help you to obtain any equipment that you may need.

ABC would like to receive a finished report on or before June 30, and I would like to receive a review draft no later than June 24. Please let me know if you need help rearranging your other work to accommodate this schedule.

Thanks and best regards,

Sheldon Jeter
President, S & J Mechanical Corp.

Introduction:
Context is
established by
restating
the previous
email

Discussion:
Survey
schedule and
completion
date

Closing:
Delivery
dates are
defined

For clarity,
it is useful
to reproduce
the mail that
you are
answering

To: **Sheldon Jeter**
From: **Jeff Donnell**
Date: **May 22, 2011**
Subject: **Re: Environmental Assessment of ABC Auto Repair facility in**
           St. Louis County

Sheldon,

Per your request, Mike and I have contacted Richard, and we have arranged to meet him at the facility on May 28 to begin the assessment. Based on Richard's description of the facility, the assessment should be completed that day unless further soil borings are required. We will contact you on the afternoon of the 28th if we identify areas that require further attention from us.

We will give you a verbal update when return on the 29th, and we will have a formal report for you by June 5.

Regards,

Jeff

Jeff Donnell
Project Manager, S&J Mechanical Corp.

Sheldon Jeter wrote:

> The purpose of this e-mail is to document a discussion that we held earlier today. Our company has been hired by ABC Auto Repair to assess their St. Louis County facility for soil contamination prior to its sale to a third party. We will meet ABC's needs by conducting a Phase I Environmental Site Assessment and a Limited Subsurface Investigation at this facility. You two are to perform this work.
>
> The facility is no longer in use. ABC asserts that the facility has one in-ground hydraulic lift and that two underground gasoline storage tanks were removed in 1994. The subsurface investigation should include soil borings near that in-ground lift and at the site of those underground tanks.
>
> The property is at 15100 Manchester Road in Ballwin, Missouri. Your contact at ABC is Richard Muller. Contact him at ABC's headquarters, 1275 Jefferson Avenue in St. Louis. His phone is 876-123-4567. Richard will give you access to the facility, will show you where the underground storage tanks were removed and will help you to obtain any equipment that you may need.
>
> ABC would like to receive a finished report on or before June 30, and I would like to receive a review draft no later than June 24. Please let me know if you need help rearranging your other work to accommodate this schedule.
>
> Thanks and best regards,
>
> ------
> Sheldon Jeter
> President, S & J Mechanical Corp.

INDEX